Cambridge Primary

Mathematics

Second Edition

Workbook 6

Steph King
Josh Lury
Series editors:
Mike Askew
Paul Broadbent

Boost

HODDER EDUCATION
AN HACHETTE UK COMPANY

Acknowledgements

The Publishers would like to thank the following for permission to reproduce copyright material.

Photo credits

p. 9 *tl*, cr, **p. 15** *tl*, *cr*, **p. 21** *tl*, *cr*, **p. 25** *tl*, *cr*, **p. 31** *tl*, *cr*, **p. 34** *tl*, *cr*, **p. 39** *tl*, *cr*, **p. 44** *tl*, *cr*, **p. 51** *tl*, *cr*, **p. 54** *tl*, *cr*, **p. 59** *tl*, *cr*, **p. 63** *tl*, *cr*, **p. 68** *tl*, *cr*, **p. 73** *tl*, *cr*, **p. 81** *tl*, *cr*, **p. 87** *tl*, *cr*, **p. 92** *tl*, *cr*, **p. 96** *tl*, *cr* © Stocker Team/Adobe Stock Photo.

t = top, *b* = bottom, *l* = left, *r* = right, *c* = centre

Every effort has been made to trace all copyright holders, but if any have been inadvertently overlooked, the Publishers will be pleased to make the necessary arrangements at the first opportunity.

Hachette UK's policy is to use papers that are natural, renewable and recyclable products and made from wood grown in well-managed forests and other controlled sources. The logging and manufacturing processes are expected to conform to the environmental regulations of the country of origin.

Orders: please contact Hachette UK Distribution, Hely Hutchinson Centre, Milton Road, Didcot, Oxfordshire, OX11 7HH. Telephone: +44 (0)1235 827827. Email education@hachette.co.uk Lines are open from 9 a.m. to 5 p.m., Monday to Friday. You can also order through our website: www.hoddereducation.com

ISBN: 978 1 3983 0124 5

First published in 2017

This edition published in 2021 by

Hodder Education,

An Hachette UK Company

Carmelite House

50 Victoria Embankment

London EC4Y 0DZ

www.hoddereducation.com

Impression number 10 9 8 7 6 5 4 3 2 1

Year 2025 2024 2023 2022 2021

Cover illustration by Lisa Hunt, The Bright Agency

Illustrations by: Alex van Houwelingen, Ammie Miske, James Hearne, Jeanne du Plessis, Natalie and Tamsin Hinrichsen, Stéphan Theron, Steve Evans, Tina Nel, Vian Oelofsen

Typeset in FS Albert 12/14 by IO Publishing CC

Printed in the UK

A catalogue record for this title is available from the British Library.

Contents

Unit 1 Number

Remember: When you see this star ★, it is showing you that the activity develops your Thinking and Working Mathematically skills!

Can you remember?

Which of the seven numbers will you position on the shaded part of the number line? Circle them.

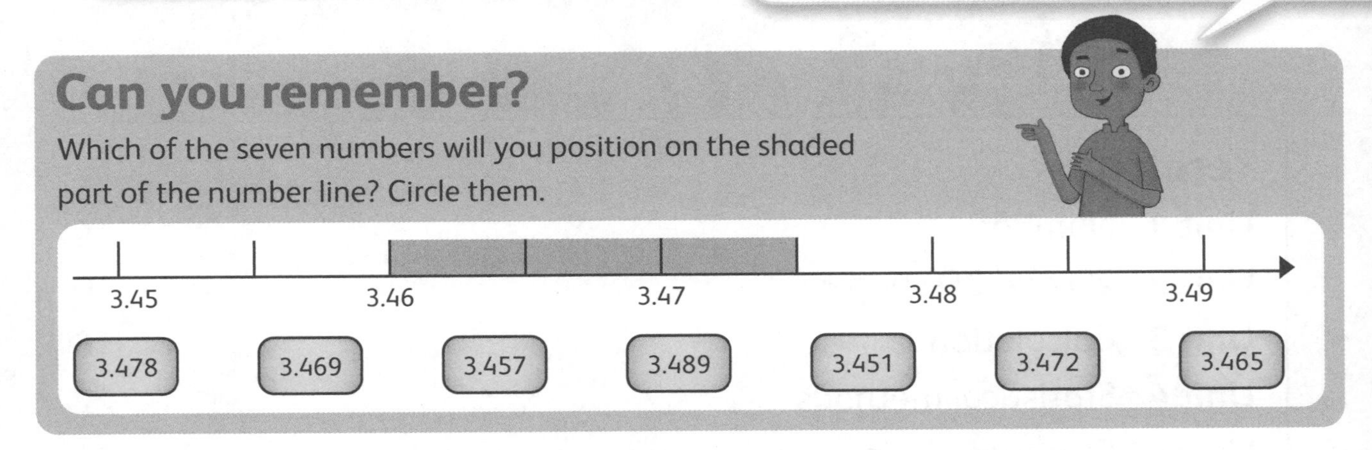

Composing, decomposing and regrouping whole numbers and decimals

1 Join the numbers with the correct digit value. Look at the example first.

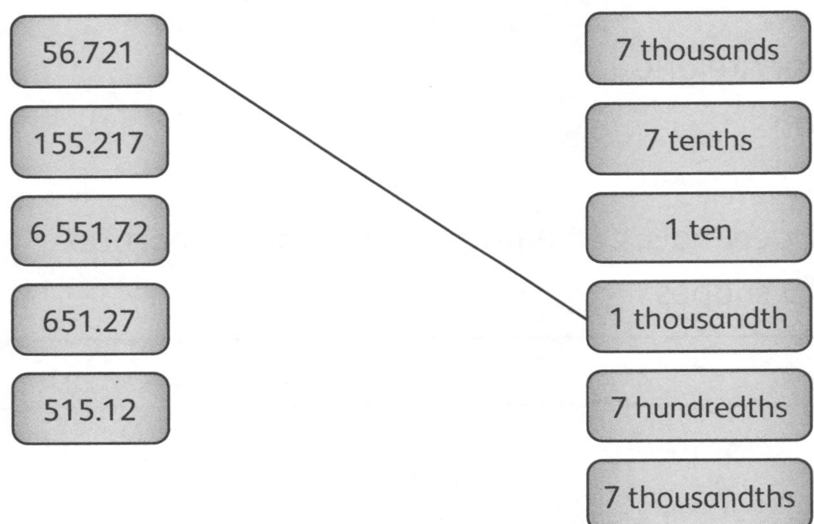

2 Fill in the missing numbers.

a 79.423 = 79 + ☐

b 79.423 = 60 + ☐ + 0.023

c 79.423 = 79 + 0.4 + ☐

d 79.423 = ☐ tenths and ☐ thousandths

3 Use these place value parts to compose the three numbers below.

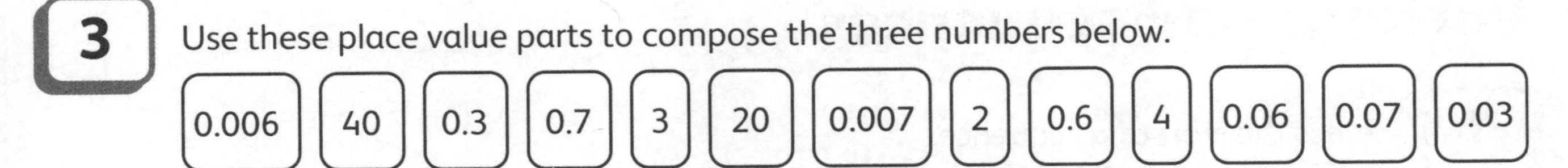

a Compose **24.76**. Colour the place value parts you need in yellow.
b Compose **42.376**. Colour the place value parts you need in red.
c Compose **3.637**. Colour the place value parts you need in green.

Multiplying and dividing whole numbers and decimals by 10, 100 and 1 000

1 Take 672.45 through the flowchart. Carry out the multiplications and divisions.

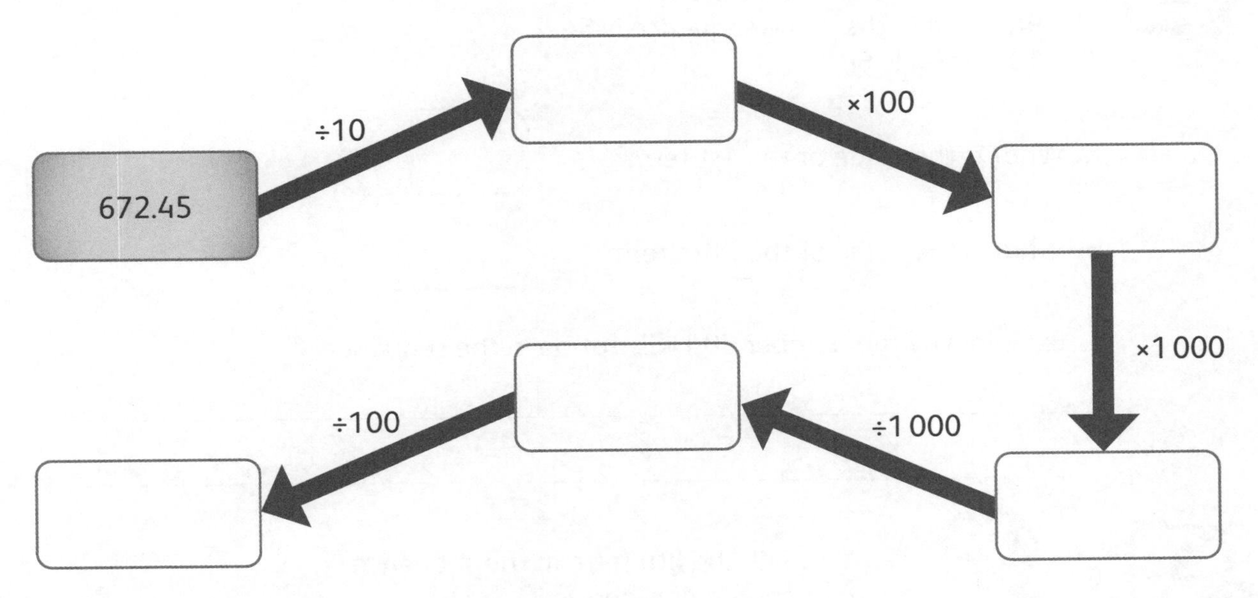

2 The distance between city A and city B is 199.96 km.
The distance between city C and city D is 100 times as far.
What is the distance between city C and city D? Show your working below.

Patterns and sequences

1 Here is the start of a sequence.

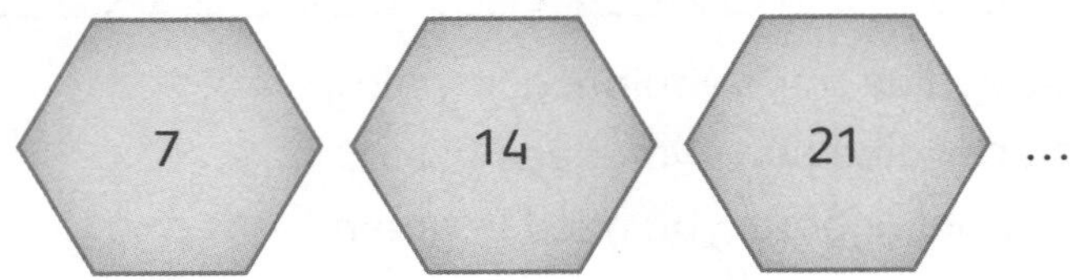

Fill in the table to show the value of the other terms in the sequence.

Position	1	2	3	6	10	19	21	42	50
Term	7	14							

2 The 4th term in a sequence is 24.
The 9th term in the same sequence is 54.
The 10th term is 60.

a What is the value of the 1st term?

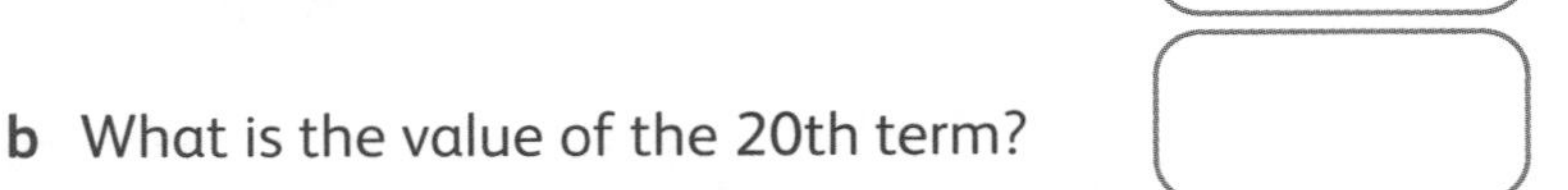

b What is the value of the 20th term?

c Explain why the number 604 will **not** be in the sequence.

3

I think the 8th term in the sequence of square numbers is 81.

Do you agree?
What can you draw or write to **convince** others of your decision?

Common multiples and common factors

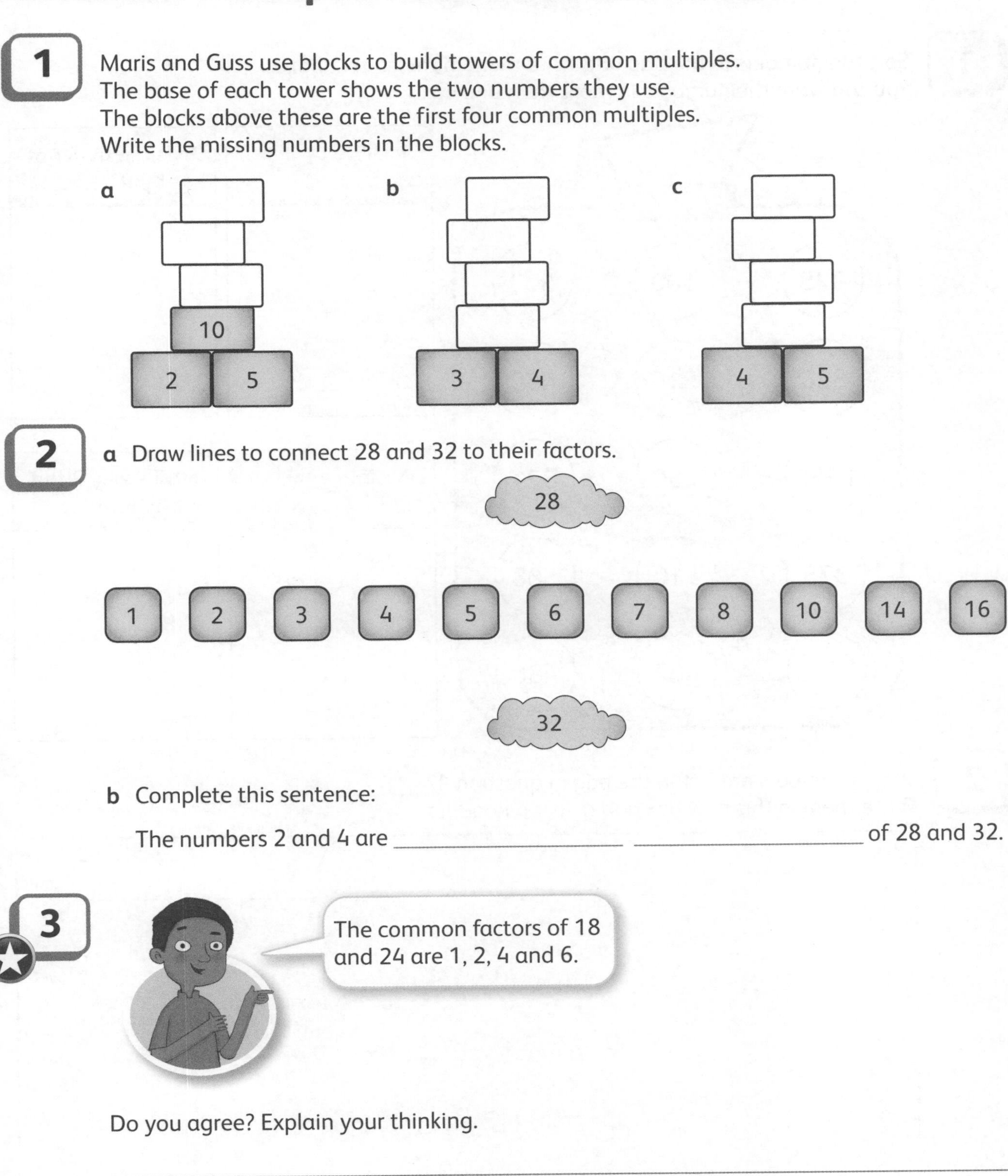

1 Maris and Guss use blocks to build towers of common multiples.
The base of each tower shows the two numbers they use.
The blocks above these are the first four common multiples.
Write the missing numbers in the blocks.

2 **a** Draw lines to connect 28 and 32 to their factors.

b Complete this sentence:

The numbers 2 and 4 are ____________ ____________ of 28 and 32.

3

Do you agree? Explain your thinking.

____

____

____

Tests of divisibility

Sort the numbers in the bag and write them in the correct boxes.
You can write the numbers in more than one box.

325	600	573
828	216	459
1 281	650	321
375	510	588
1 143	402	486

Divisible by 3, 6 and 9	Divisible by 3 but not by 9

Divisible by 9 but not by 6	Divisible by 3 but not by 6 or 9

2 Which numbers are left in the bag in question **1**?
Write them in this new box and give it a heading.

Unit 1 Number

Self-check

 I can do this.

 I can do this, but I need to keep trying.

 I can't do this yet.

	What can I do?	😉	😐	☹
1	I can state and explain the value of each digit in decimals (tenths, hundredths and thousandths).			
2	I can compose and decompose numbers using place value, including decimals (tenths, hundredths and thousandths).			
3	I can regroup numbers, including decimals (tenths, hundredths and thousandths) in a variety of ways, for example: 9.234 = 4 + 5.2 + 0.034 or 8.004 + 1.23			
4	I can multiply and divide whole numbers and decimals by 10, 100 and 1 000.			
5	I can identify missing numbers in linear sequences, for example: 4, ____, ____, 16, …			
6	For simple linear sequences, I can find, for example, the 10th, 31st or 99th terms without listing all the terms.			
7	I can explain that a common multiple is number that is a multiple of two or more numbers.			
8	I can list at least three common multiples of two numbers.			
9	I can explain that a common factor is a number which is a factor of two or more numbers.			
10	I can list all the common factors of two numbers.			
11	I can say whether or not a number is divisible by 3, 6 or 9.			
12	I can fill in digits to make a number divisible by 3, 6 or 9, for example: Put a digit in the box to make a number divisible by 9: 23[3]1.			

I need more help with:

____

____

____

Unit 2 2D and 3D shapes

Can you remember?

Label each angle: **acute** or **obtuse**.

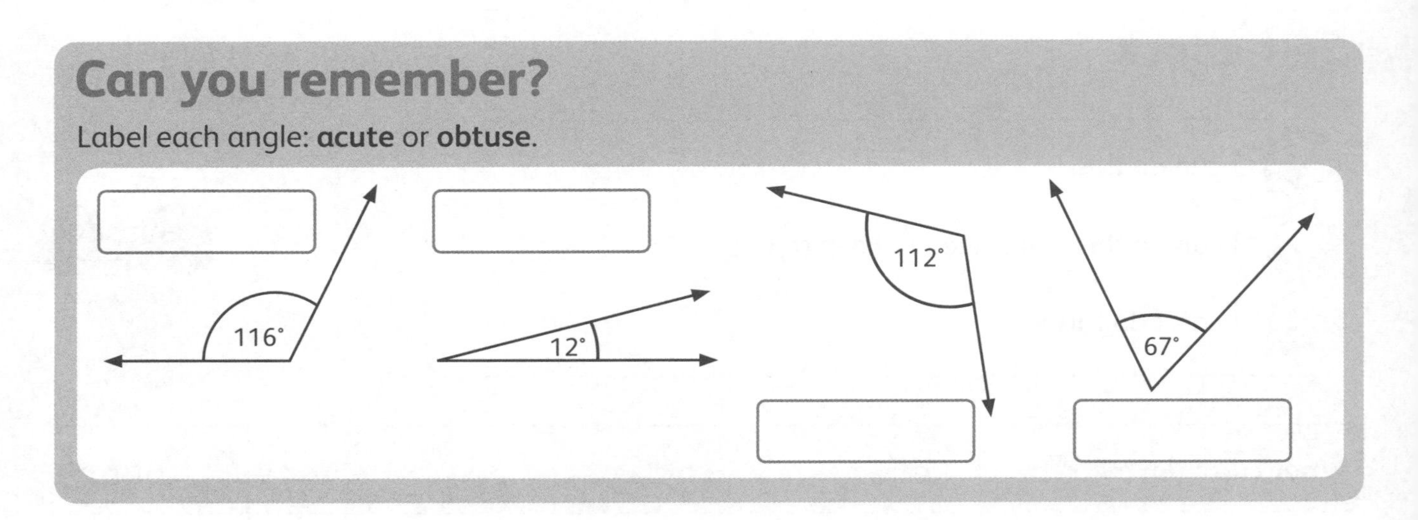

Measuring and drawing angles

1 Measure each angle accurately. Write the measurement at the angle.

Join dots to form angles. Then measure and label each angle that you have drawn. Look at the example.

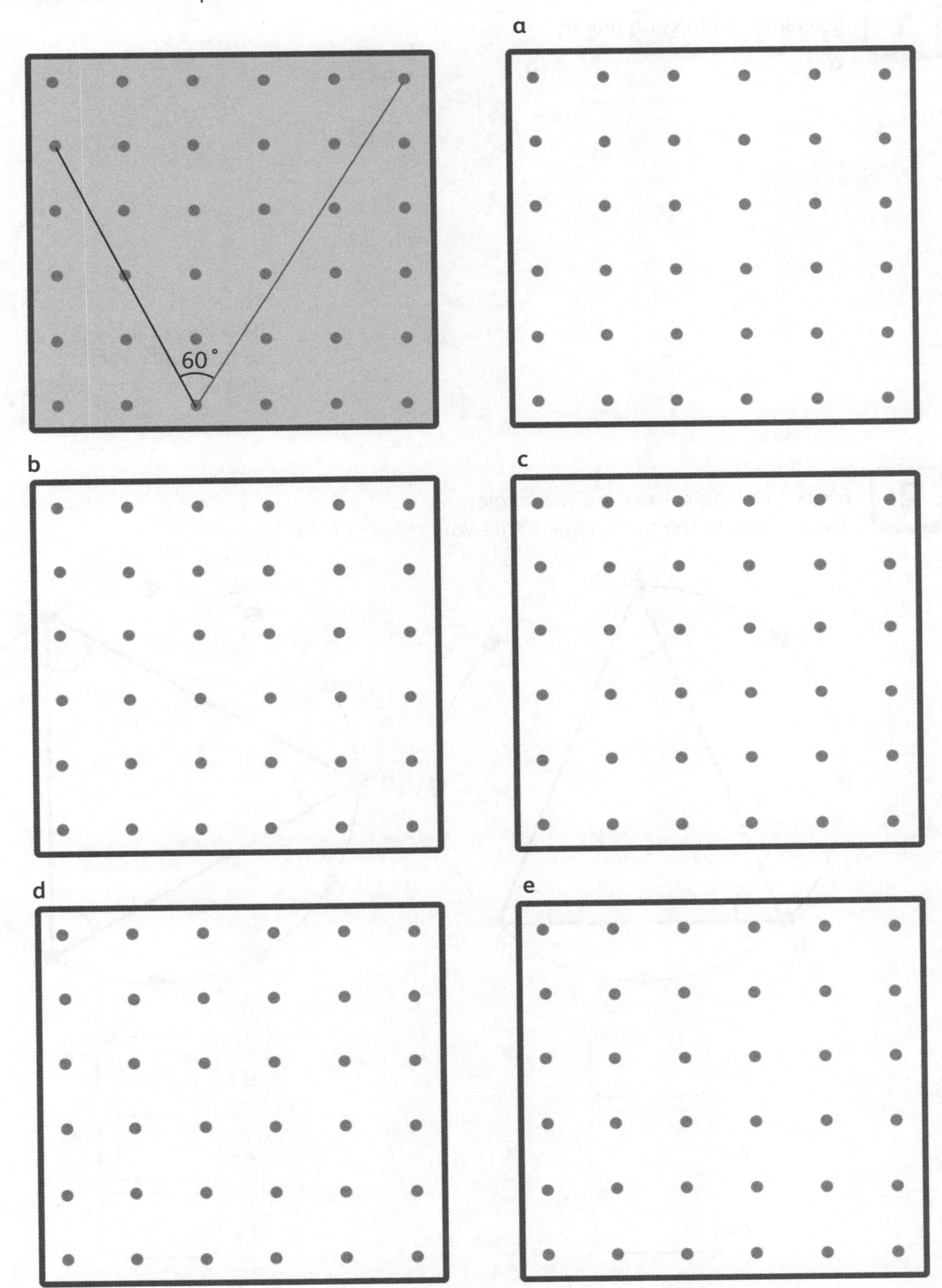

Calculating angles in triangles

1 Calculate the missing angles.

a

70° 40° ?

°

b

90° 60° ?

°

c

? 140° ?

This is an isosceles triangle, so both unknown angles are: °

2 For each triangle, measure two angles.
Then calculate the third angle. Write your results below.

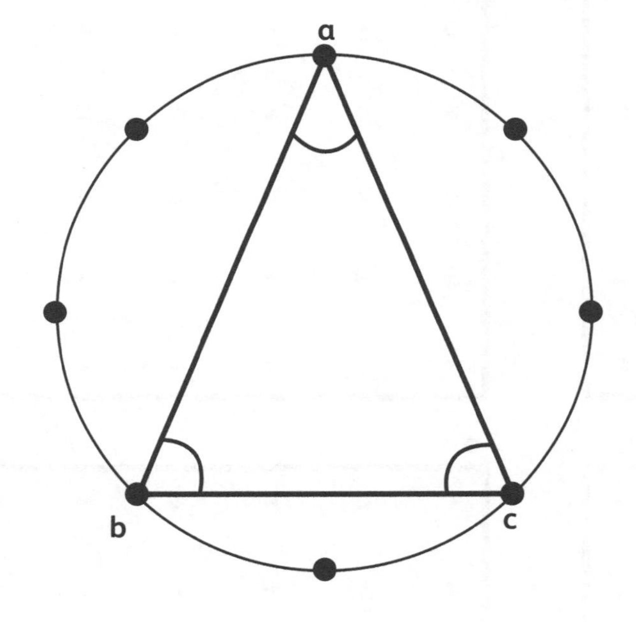

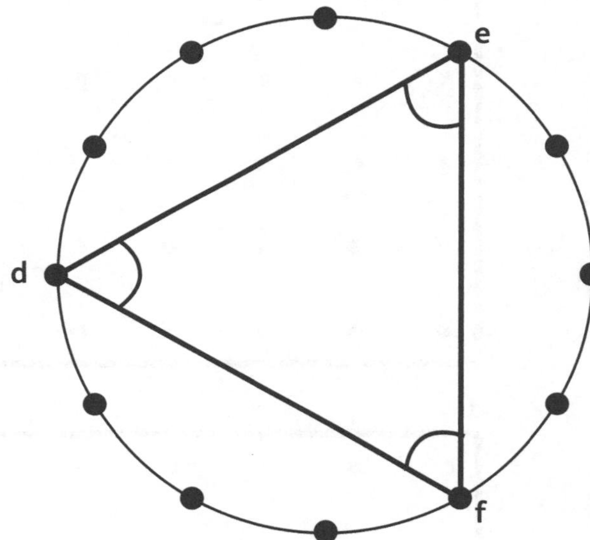

a = °

b = °

c = °

d = °

e = °

f = °

Join three dots in each circle to make four different triangles.
Measure two angles each time. Calculate the third angle. Label each angle.
Then write the type of triangle: isosceles, equilateral or scalene.

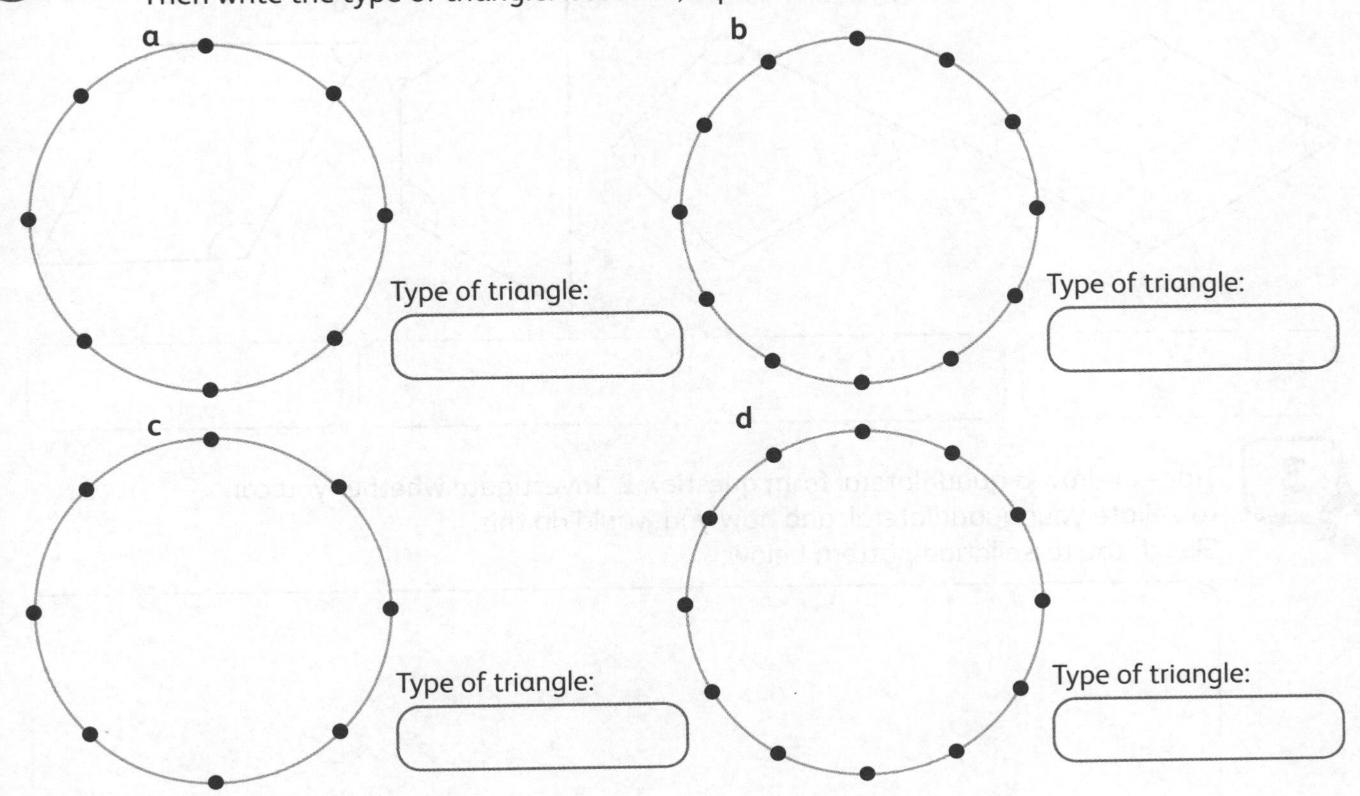

c

Type of triangle:

d

Type of triangle:

Properties of quadrilaterals

1 Draw the shape for each description. Write its name. Mark any parallel lines.

a A quadrilateral with one pair of parallel lines

____

b A quadrilateral with two pairs of parallel lines

____

c A quadrilateral with no parallel lines

____

d A quadrilateral with four sides of the same length

____

e A quadrilateral with all sides of different lengths

____

f A quadrilateral with two right angles

____

2 Measure and label the angles in each quadrilateral. Mark any lines of symmetry. Write the name of each shape under it.

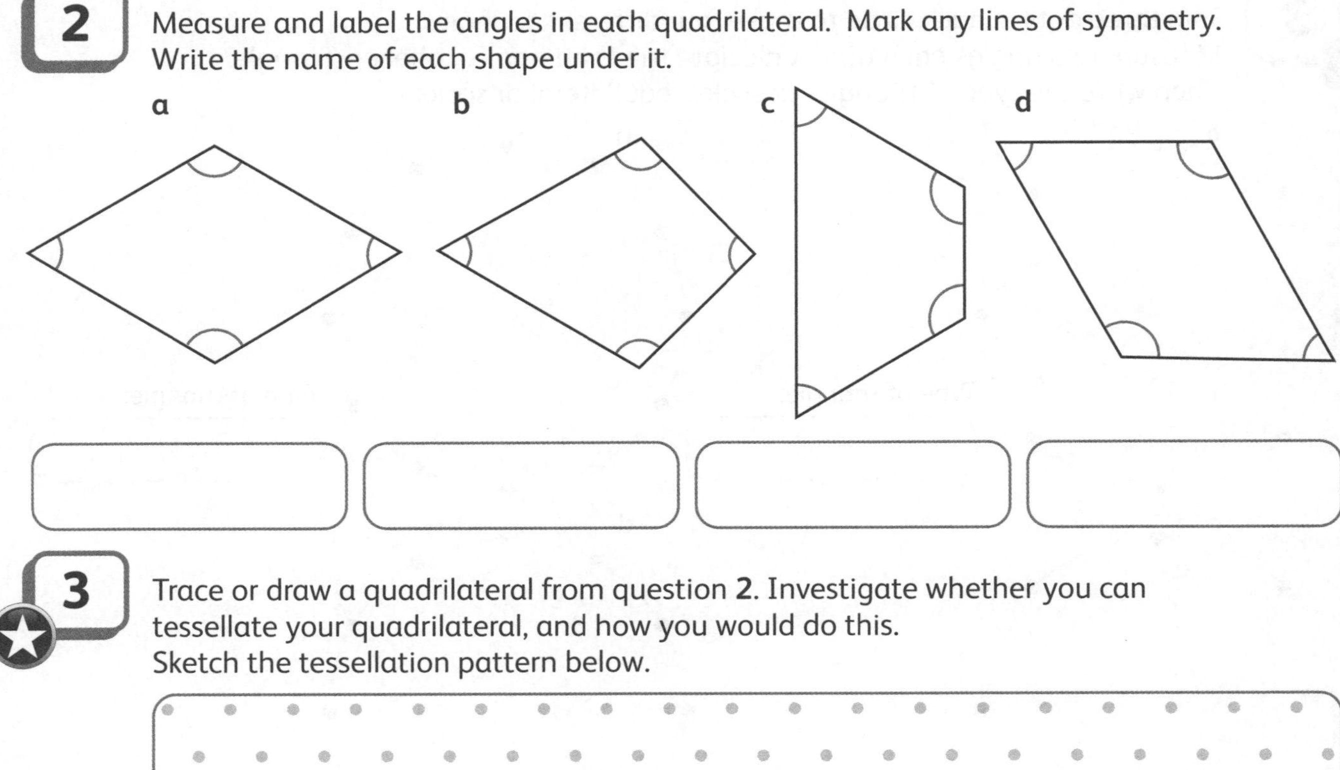

3 Trace or draw a quadrilateral from question **2**. Investigate whether you can tessellate your quadrilateral, and how you would do this.
Sketch the tessellation pattern below.

Unit 2 2D and 3D shapes

Self-check

 I can do this.

 I can do this, but I need to keep trying.

 I can't do this yet.

What can I do?		:-)	:-|	:-(
1	I can identify, describe and classify quadrilaterals, including reference to angles, symmetrical properties, parallel sides and diagonals.			
2	I can sketch quadrilaterals, given certain properties.			
3	I can classify angles.			
4	I can estimate and measure angles.			
5	I can draw triangles.			
6	I can calculate missing angles in a triangle.			
7	I can show how any quadrilateral can tessellate.			

I need more help with:

____

____

____

Unit 3 Calculation

Can you remember?

Complete these calculations.

a	b	c
123 × 10 = ☐	321 ÷ 100 = ☐	☐ ÷ 100 = 0.555
123 ÷ 10 = ☐	321 × 100 = ☐	☐ ÷ 100 = 5.55
☐ × 10 = 123	321 ÷ ☐ = 0.321	☐ ÷ 100 = 55.5

Addition and subtraction with positive and negative numbers

1

a The table shows the difference in summer and winter temperatures in different locations. Fill in the missing information.

Location	A	B	C	D	E
Summer	28 °C	30 °C	15 °C		10 °C
Winter	−2 °C	−4 °C		−8 °C	
Difference			**18°**	**27°**	**22°**

b The winter temperature at midday in location F is −3 °C.
By midnight, the temperature is −18 °C.
What is the difference between the temperature at midday and the temperature at midnight?

In location F, the difference between the temperature at midday and the temperature at midnight is ___________°.

2 Complete these calculations.

a $-34 + 36 =$ ☐

$-24 + 26 =$ ☐

$-14 + 16 =$ ☐

$-4 + 6 =$ ☐

b $22 - 46 =$ ☐

$32 - 66 =$ ☐

$42 - 86 =$ ☐

$52 - 106 =$ ☐

c ☐ $+ 25 = 9$

$9 - 25 =$ ☐

$20 -$ ☐ $= -80$

$-80 +$ ☐ $= 20$

Describe any patterns you notice in parts **a** and **b**.

____

____

Addition and subtraction

1 Are these statements true or false? Tick (✓) or make a cross (×). Correct any that are false.

	Statement	True (✓)	False (×)	Correction
a	3 920 + 2 780 > 7 000			
b	6 815 + 3 789 > 9 500			
c	3 224 + 7 018 > 10 000			
d	3 924 − 2 783 < 2 000			
e	6 911 − 3 784 < 3 200			
f	5 958 − 3 091 < 3 000			

2 Solve these additions and subtractions.
Show your working to demonstrate the method you used.

a 1 200 + 900	**b** 2 315 − 1 030	**c** 6 945 + 1 997
d 4 732 + 3 675	**e** 3 655 − 1 255	**f** 7 027 − 6 955

3 The total mass of three sacks of vegetables is 18 475 g.
Sack A is double the mass of sack B.
Sack C has a mass that rounds to 7 kg to the nearest 1 000 g.
Find three possible masses for sacks A, B and C.

	Solution 1	Solution 2	Solution 3
Sack A			
Sack B			
Sack C			

Using letters to represent quantities

1 Pia has ***f*** cents in her moneybox. She spends 20 cents on a lollipop. Pia has ***g*** cents left.

a Represent the problem as a calculation.

____

b Find at least five possible values for each of ***f*** and ***g***.

f								
g								

2 The children arrange ten boxes of toys on three shelves.

a **Critique** Jin's idea. How will you explain any mistake that he has made?

____

____

b Now represent the problem using letters of your choice to label each shelf.
Find five possible solutions for the number of boxes that the children can place on each shelf.

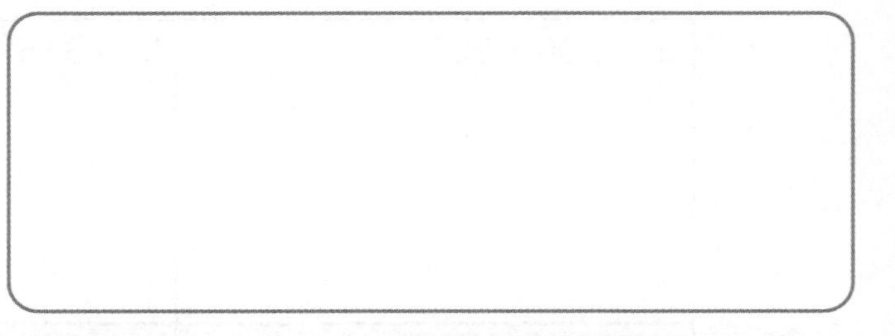

Shelf	Solution 1	Solution 2	Solution 3	Solution 4	Solution 5

Simplifying calculations

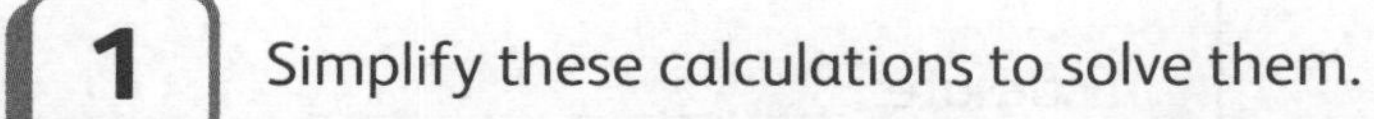

1 Simplify these calculations to solve them.

a $35 \times 6 + 4 \times 35$	**b** $72 \times 12 - 2 \times 72$	**c** $40 + 96 \div 3 + 60$

2 **a** Use four of the five number cards to make all three calculations correct.

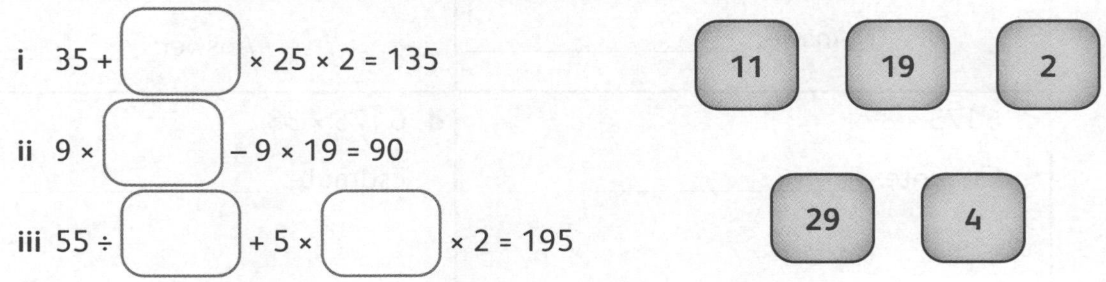

i $35 + \square \times 25 \times 2 = 135$

ii $9 \times \square - 9 \times 19 = 90$

iii $55 \div \square + 5 \times \square \times 2 = 195$

11 19 2 29 4

b Show how you would simplify each calculation to make it easier to solve.

i ____________________________________

ii ____________________________________

iii ____________________________________

Multiplying whole numbers up to 10 000

1 Jin multiplies a whole number by 25. He gets a 5-digit number.
Elok multiplies a larger whole number by 25. She also gets a 5-digit number.
The difference between their answers is a multiple of 100.

a What is the smallest number that each child can start with? Jin ☐ Elok ☐

b What is the largest number that each child can start with? Jin ☐ Elok ☐

c What number can each child start with so that the answers have the largest possible difference? Jin ☐ Elok ☐

2 Complete these calculations. Make an estimate first.

a 5 329 × 6 Estimate: ______ Answer: ______	**b** 5 329 × 26 Estimate: ______ Answer: ______
c 6 175 × 8 Estimate: ______ Answer: ______	**d** 6 175 × 38 Estimate: ______ Answer: ______

3 The sides in this design follow a pattern.
The lengths of the sides of each square become 12 mm longer each time.
The length of the sides of square A is 36 mm.

a What is the area of each square?

A = [ ] mm^2

B = [ ] mm^2

C = [ ] mm^2

D = [ ] mm^2

E = [ ] mm^2

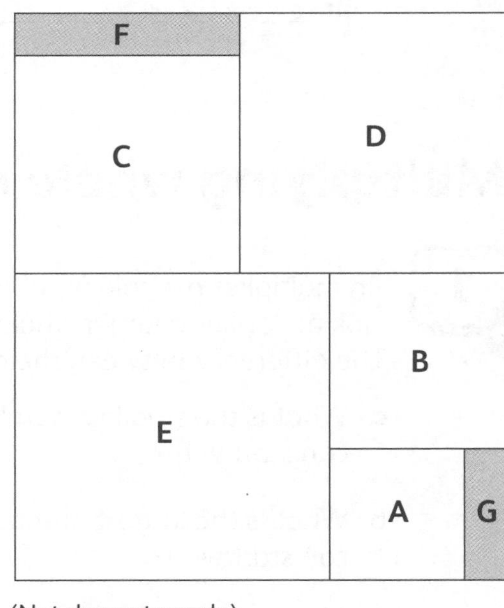

(Not drawn to scale)

b Work out the area of the shaded shapes. F = ______ G = ______

Unit 3 Calculation

Self-check

See how much you know!

 I can do this.

 I can do this, but I need to keep trying.

 I can't do this yet.

What can I do?			
1 I can add a positive integer to a negative integer, for example: –12 + 40.			
2 I can subtract a positive integer from an integer where the answer is negative, for example: 22 – 35.			
3 I can find the difference between positive and negative integers, and between two negative integers, for example: 30 – 5 = 25, –30 – 5 = –35, –30 + 5 = –25, 30 + 5 = 35.			
4 I can estimate, add and subtract integers, including negative numbers.			
5 I can solve problems involving addition and subtraction.			
6 I can use letters to represent quantities that vary in addition and subtraction calculations.			
7 I can compose, decompose and regroup integers, in order to simplify calculations.			
8 I can apply the laws of arithmetic and the order of operations to simplify calculations, for example: 25 × 3 × 4 + 5 = 25 × 4 × 3 + 5 =100 × 3 + 5 = 300 + 5 = 305.			
9 I can estimate the product of a whole number up to 10 000 with 1-digit or 2-digit whole numbers.			
10 I can multiply whole numbers up to 10 000 by 1-digit or 2-digit whole numbers.			

I need more help with:

____

____

____

Unit 4 Statistical methods

Can you remember?

List the different types of charts you can remember.

- ______________________ • ______________________
- ______________________ • ______________________
- ______________________ • ______________________

Comparing data and charts

1 Investigate the opinions (views) of your classmates or family about the question:

What is the best way to be healthy?

I will find out how many minutes of exercise each person does.

Consider factors such as:
- diet
- exercise
- socialisation (mixing with others)
- hygiene.

Decide on and write your own question to investigate how to stay healthy.

____

____

2 Collect information from your classmates or family about their opinions on the question you chose to investigate. Record the results accurately.

3 Present the information using a chart of your choice.

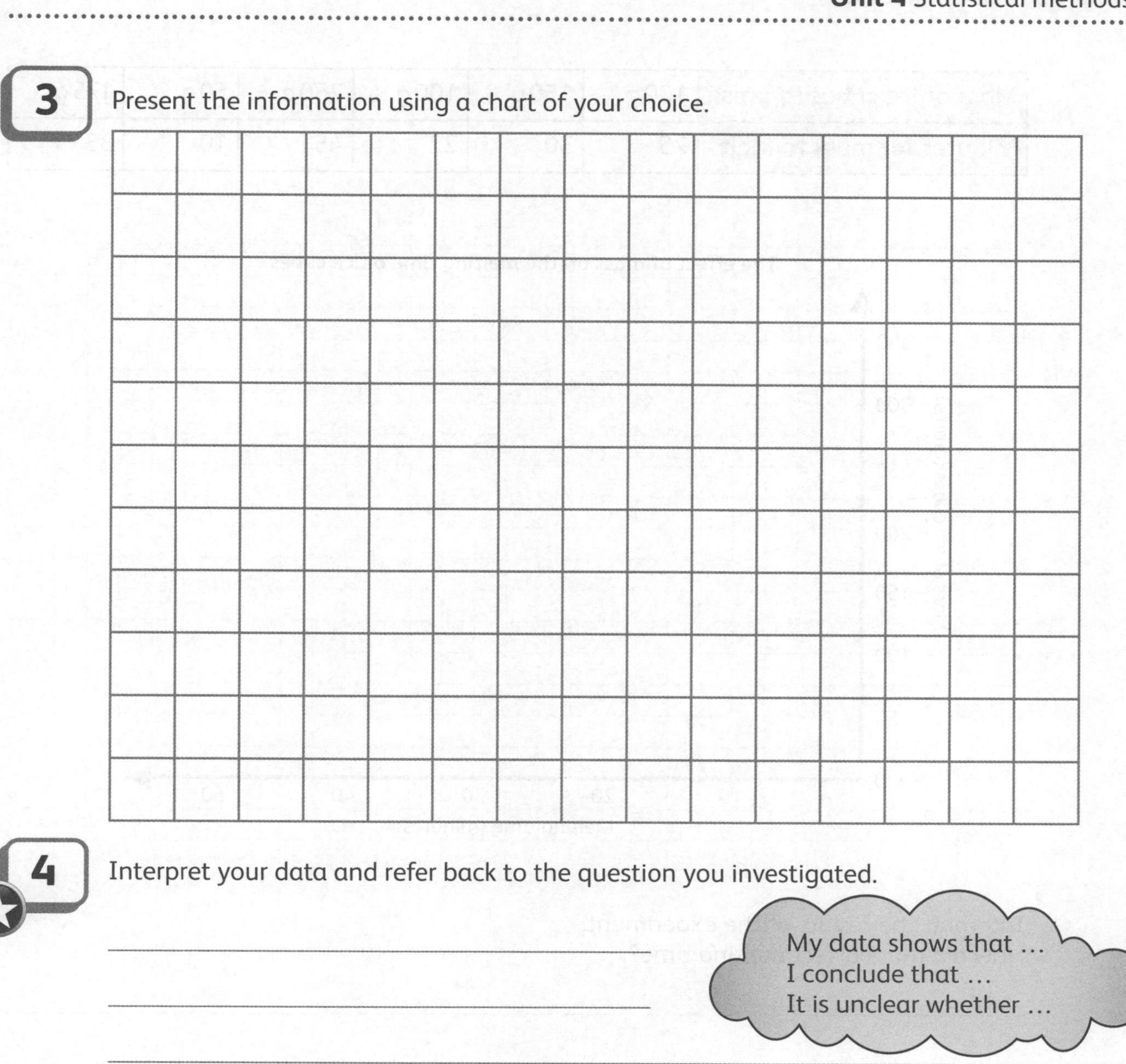

4 Interpret your data and refer back to the question you investigated.

My data shows that …
I conclude that …
It is unclear whether …

Line graphs and scatter plots

1 Sanchia completed a science experiment to investigate the question:

How does mass affect the melting time of ice cubes?

Sanchia measured the mass of some ice cubes.
Then she measured how long each mass of ice cubes took to melt.
Record Sanchia's results in a scatter plot on the next page.

Mass of ice cubes (grams)	100g	150g	100g	250g	50g	175g
Minutes for mass to melt	25	30	22	45	10	35

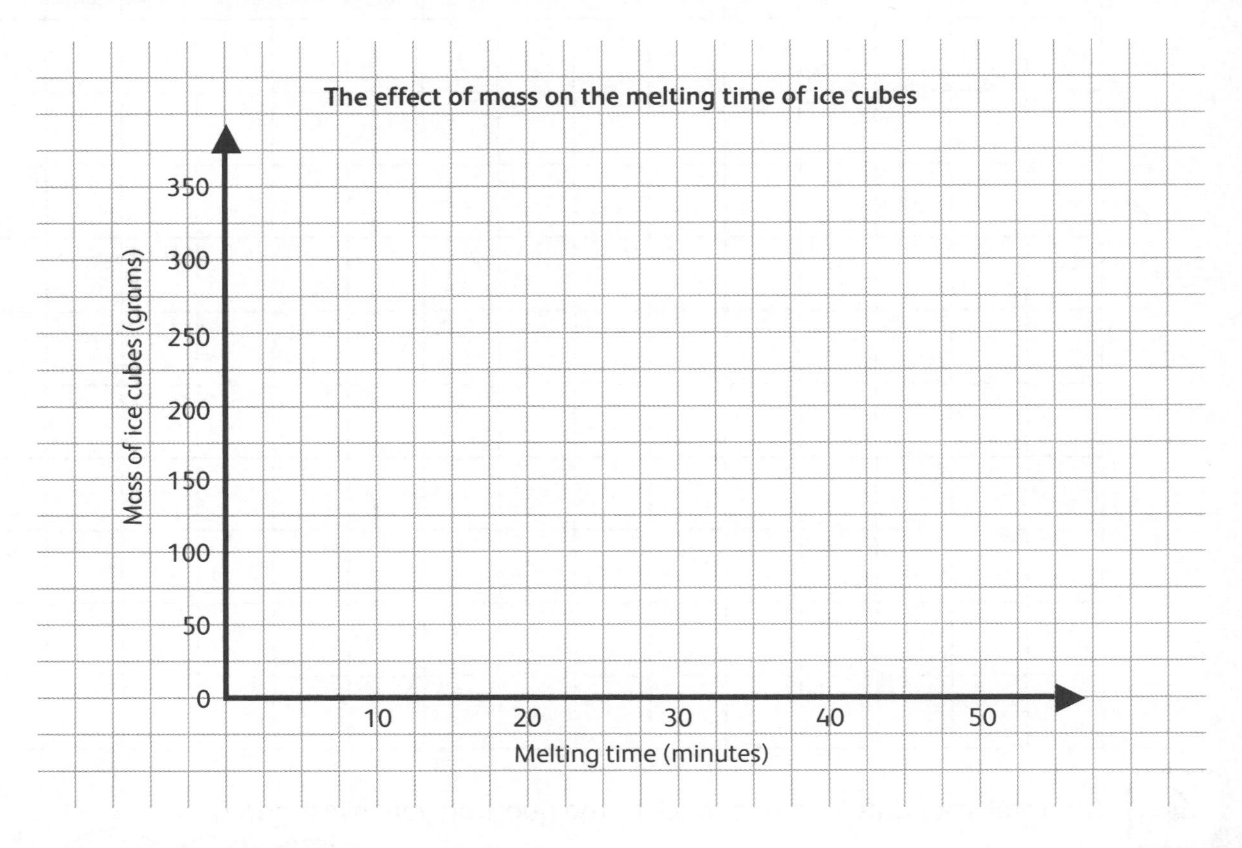

Interpret the results of the experiment.
How did mass affect melting time?

____

____

____

____

2 Use your findings and your scatter plot from question **1** to predict:

a The time it took for 125g of ice cubes to melt in Sanchia's experiment

____

b The time it took for 40g of ice cubes to melt

____

c The mass of ice cubes that took 27 minutes to melt

____

Unit 4 Statistical methods

Self-check

 I can do this.

 I can do this, but I need to keep trying.

 I can't do this yet.

What can I do?	I can do this.	I can do this, but I need to keep trying.	I can't do this yet.
1 I can compare and interpret two or more different representations of the same data.			
2 I can decide on what data might be helpful to collect in order to carry out an investigation.			
3 I can pose questions to investigate.			
4 I can predict anticipated (expected) outcomes of an investigation.			
5 I can record, organise and represent different kinds of data.			
6 I can interpret data and identify patterns.			

I need more help with:

____

____

____

Unit 5 Fractions, decimals, percentages, ratio and proportion

Can you remember?

Fill in the gaps to make each statement true.

a $\frac{1}{2} = \frac{\square}{4} = \frac{\square}{8} = \frac{\square}{16}$

b $\frac{1}{3} > \frac{\square}{6}$

c $\frac{3}{9} > \frac{\square}{27}$

d $\frac{1}{2} = \frac{\square}{10} = \frac{\square}{100}$

Simplifying fractions

1 Fill in the missing numbers to show how to simplify these fractions.

a

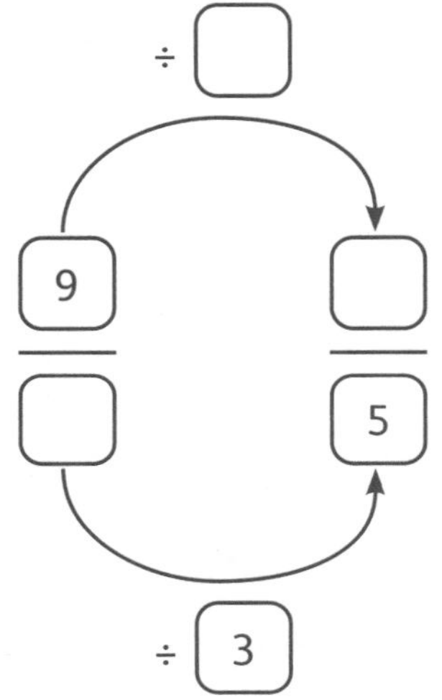

b

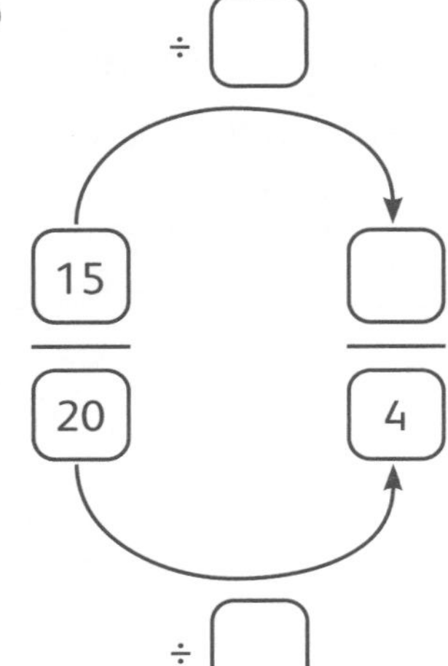

c

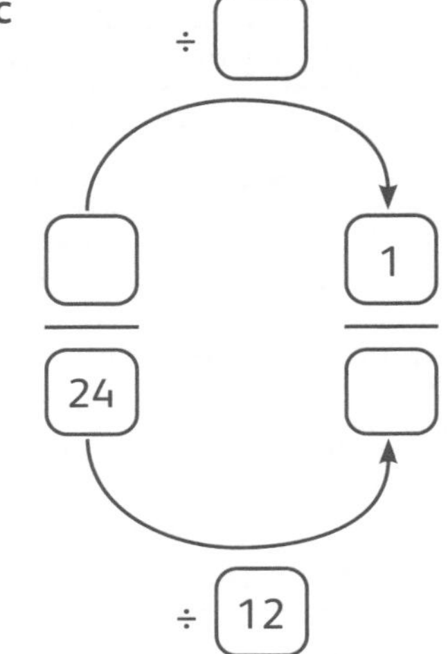

2 A bag has 25 marbles. Ten marbles are red. A box has five marbles. The fraction of red marbles in the box is equivalent to the fraction of red marbles in the bag. What fraction of the box of marbles is red? $\frac{\square}{\square}$

3 All these fractions can be simplified to fifths or tenths, but which is which? Simplify them to find out. Write any simplified fractions as decimals.

a $\frac{3}{30}$

b $\frac{20}{25}$

c $\frac{11}{55}$

d $\frac{12}{30}$ **e** $\frac{36}{40}$

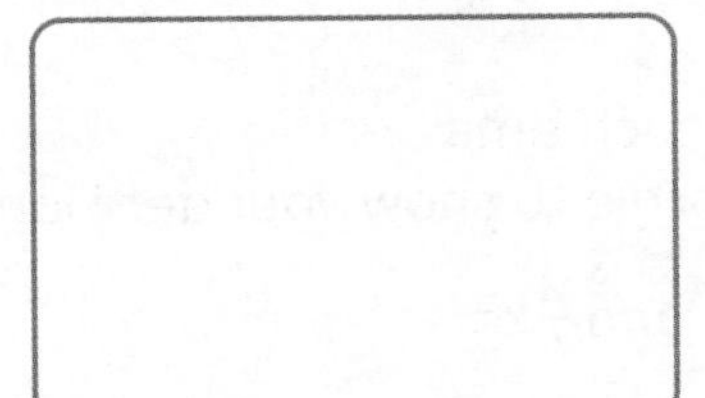
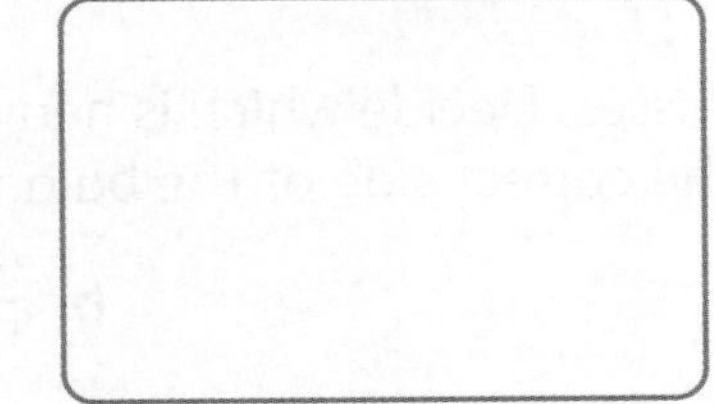

Fractions and division

a Use these shapes to represent the fraction $\frac{4}{9}$ as a division.

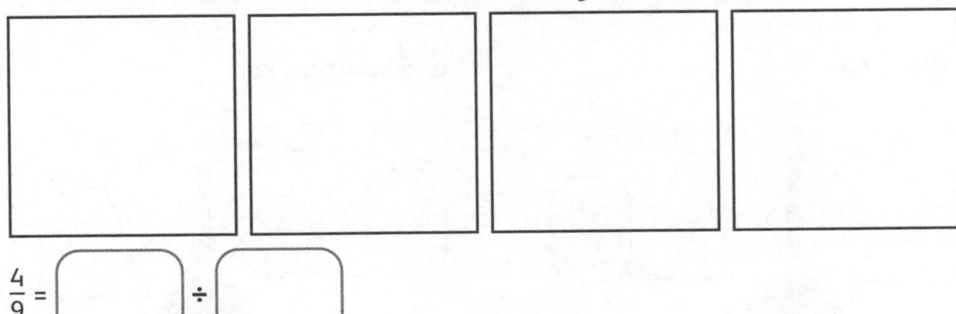

$\frac{4}{9}$ = ☐ ÷ ☐

b Use these shapes to represent the division 5 ÷ 4 as a fraction.

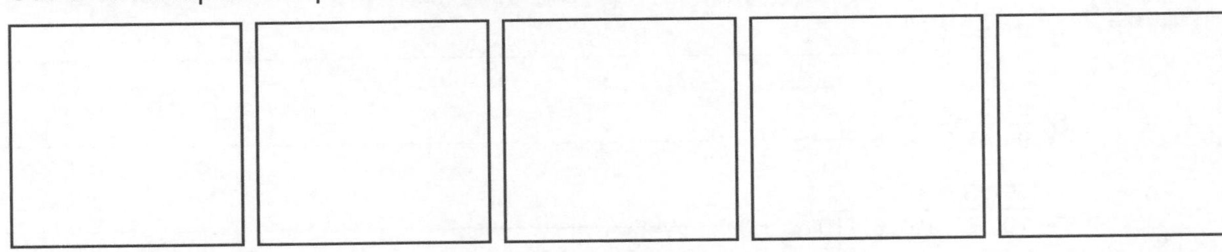

5 ÷ 4 = $\frac{☐}{☐}$

2

a A ribbon is 5 metres long. Maris cuts it into three equal pieces.

What is the length of each piece?

$\frac{☐}{☐}$ metres

b Guss divides a 7 kg sack of rice equally between four containers.

What is the mass of rice in each container?

$\frac{☐}{☐}$ kilograms

Comparing and ordering fractions

1 Look at the pairs of masses. Decide which is heavier each time.
Write the masses on the correct side of the balance scale to show your decisions.

a $\frac{3}{4}$kg and $\frac{4}{5}$kg

b $\frac{3}{10}$kg and $\frac{2}{4}$kg

c $\frac{3}{5}$kg and $\frac{2}{3}$kg

d $\frac{3}{8}$kg and $\frac{2}{5}$kg

2 Order each set of fractions from smallest to largest.

a $\frac{4}{5}, \frac{2}{3}, \frac{3}{5}, \frac{1}{3}$ __________

b $\frac{5}{9}, \frac{3}{4}, \frac{7}{9}, \frac{2}{4}$ __________

c $\frac{7}{10}, \frac{5}{6}, \frac{3}{10}, \frac{1}{6}, \frac{2}{10}$ __________

3 Elok colours in $\frac{7}{8}$ of her shape. Banko colours in $\frac{9}{10}$ of his shape.

Who has coloured in a larger fraction of their shape? __________

Fractions as operators

1 Find these fractions of amounts.

a $\frac{1}{5}$ of 30 = ☐

$\frac{3}{5}$ of 30 = ☐

$\frac{3}{5}$ of 60 = ☐

b $\frac{1}{6}$ of 42 = ☐

$\frac{5}{6}$ of 42 = ☐

$\frac{5}{6}$ of 84 = ☐

c $\frac{1}{9}$ of 54 = ☐

$\frac{5}{9}$ of 54 = ☐

$\frac{5}{9}$ of 108 = ☐

2 In a pictogram, a circle has the value 24.
Complete the calculations to show which numbers are represented here.

a

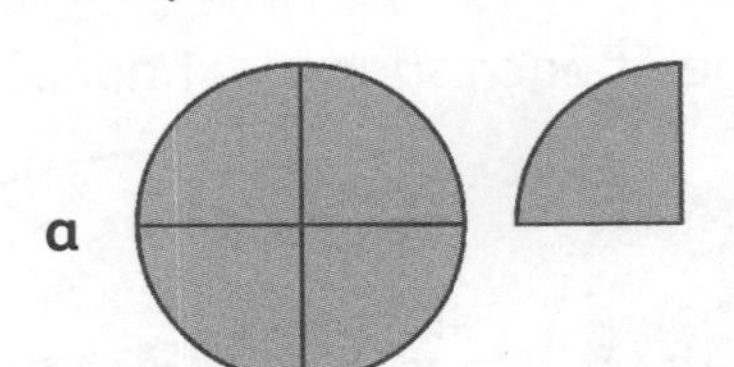

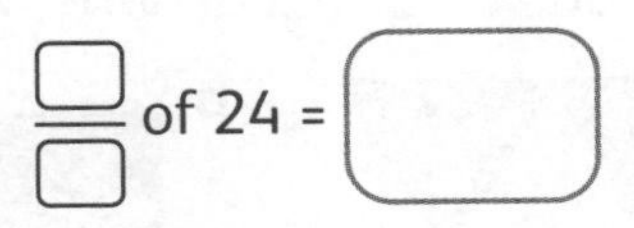

$\frac{\square}{\square}$ of 24 = ☐

b

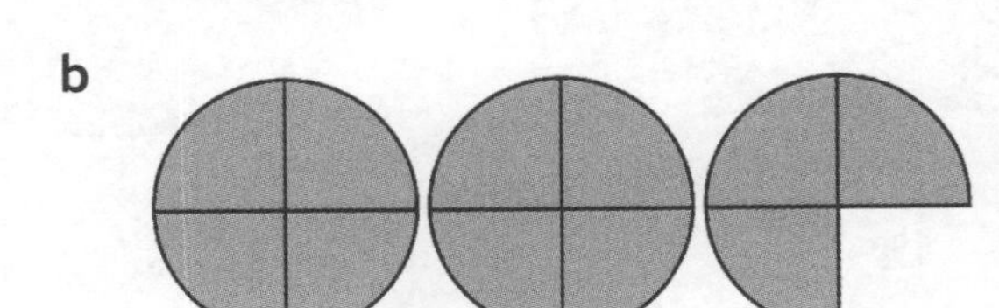

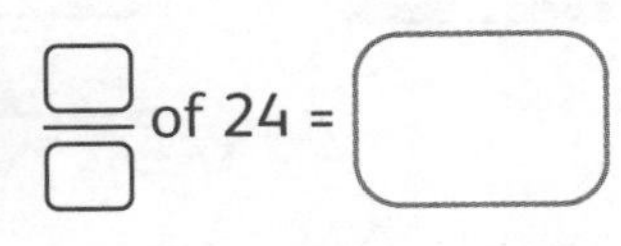

$\frac{\square}{\square}$ of 24 = ☐

3 Sanchia cycles $\frac{7}{10}$ of a 1 600 metre track.
Guss cycles $\frac{5}{3}$ of a 900 metre track.
How many more metres does Guss cycle than Sanchia? ______________ metres.

Fraction, decimal and percentage equivalences

1 Write fraction, decimal and percentage equivalents to match each diagram.

a

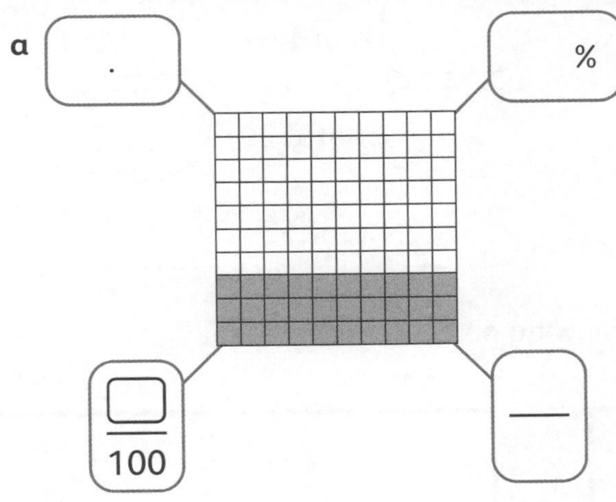

b

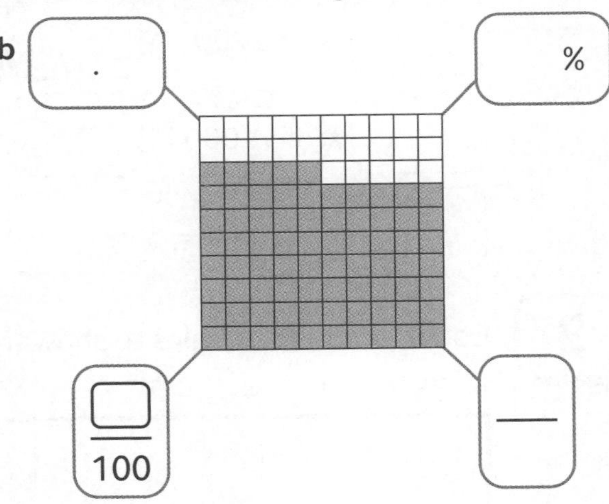

2 Draw lines to join equivalent values.

9 %	0.5	$\frac{8}{10}$	45 %	
50 %	$\frac{90}{100}$	0.09	0.45	0.8
$\frac{1}{2}$	0.9	90 %	$\frac{4}{5}$	

Percentages of shapes and quantities

1 Each shape is made up of equal parts. What percentage of each shape is shaded?

a 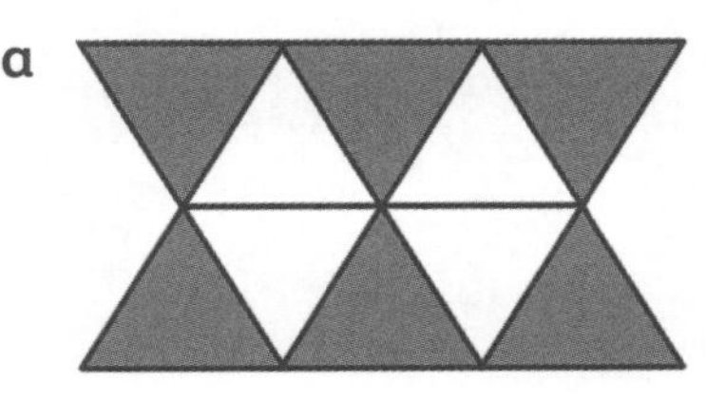

b 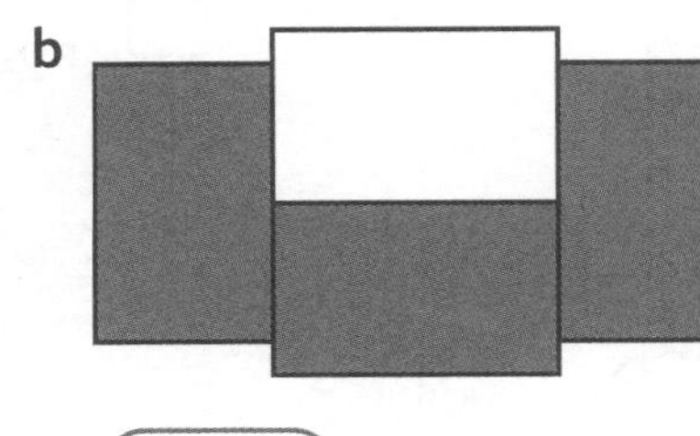

c 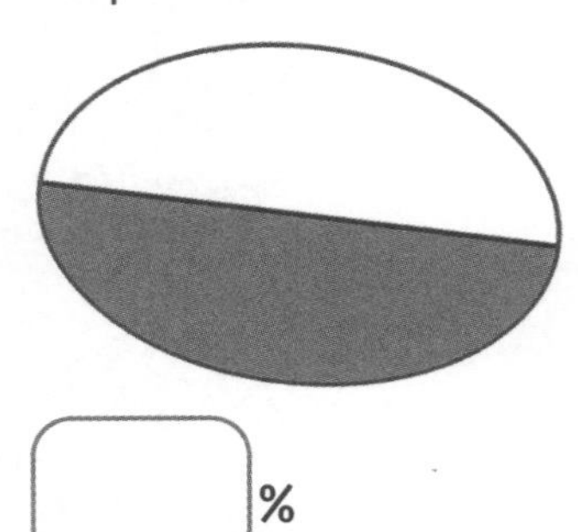

a ☐ % b ☐ % c ☐ %

2 Spin a 1 to 6 spinner. Look for the number in the table to see what percentage to find. Choose any calculation from below. Fill in the percentage and then the answer. Spin and repeat until you have completed all the calculations.

Spin a 1	Spin a 2	Spin a 3	Spin a 4	Spin a 5	Spin a 6
Find 10 %	Find 20 %	Find 25 %	Find 50 %	Find 75 %	You choose

☐ % of \$50 = \$☐ ☐ % of \$80 = \$☐

☐ % of \$200 = \$☐ ☐ % of \$10 = \$☐

☐ % of \$1 000 = \$☐ ☐ % of \$120 = \$☐

☐ % of \$40 = \$☐ ☐ % of \$5 = \$☐

3 Colour in the rectangles to show the following percentages.

a 35 %

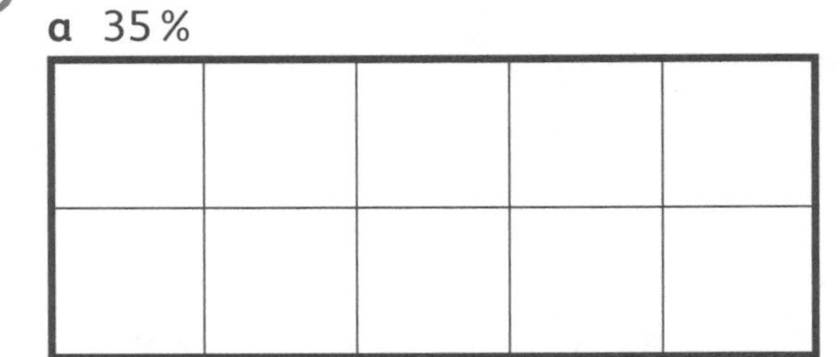

b 85 %

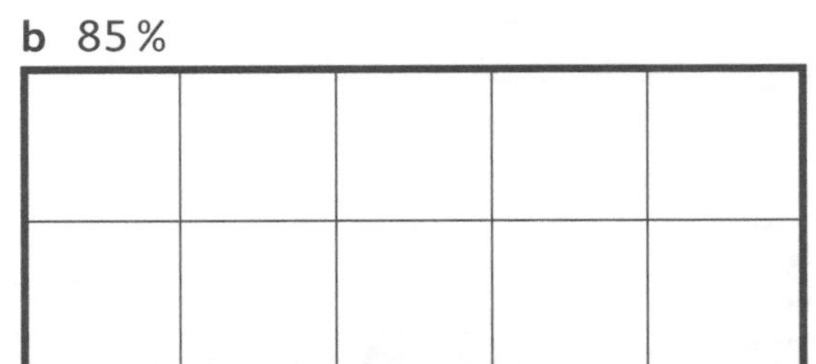

4 In a box of 240 fruits, 25 % are apples, 40 % are bananas and 5 % are oranges. The remaining fruits are pineapples. How many of each fruit are there?

☐ fruits are apples ☐ fruits are bananas

☐ fruits are oranges ☐ fruits are pineapples

Unit 5 Fractions, decimals, percentages, ratio and proportion

Self-check

See how much you know!

 I can do this.

 I can do this, but I need to keep trying.

 I can't do this yet.

What can I do?			
1 I can use my knowledge of equivalence to write fractions in their simplest form.			
2 I can represent a division problem as a fraction, knowing that the numerator is the dividend and the denominator is the divisor.			
3 I can compare and order proper fractions with different denominators.			
4 I can solve context problems involving proper and improper fractions as operators.			
5 I can solve problems involving proper and improper fractions as operators.			
6 I know and can give benchmark equivalences of fractions, decimals (one or two decimal places) and percentages.			
7 I can show on diagrams, the percentages (in multiples of 5) of shapes, for example: Shape A is 35% of a rectangle.			
8 I can calculate percentages (in multiples of 5) of quantities, for example: Find 15% of 4 kg.			

I need more help with:

____

____

____

Unit 6 Probability

Can you remember?

In the space below, draw or describe an event for each probability shown on the scale.
Connect each event to the probability scale with a line.

impossible — unlikely — equally likely — likely — certain

Probability and proportionality

1 The spinner on the right will give you a one in six chance of spinning a 1.
Write the chances of each of the following:

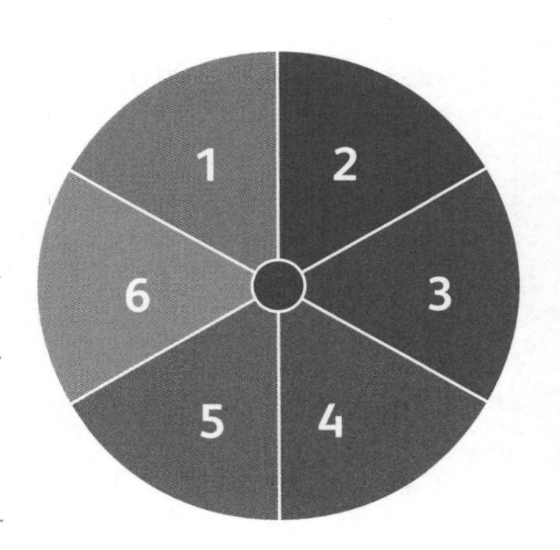

a Spinning a 6 ______________________

b Spinning a 3 ______________________

c Spinning a number greater than 1 ______________________

d Spinning a number less than 6 ______________________

e Spinning an odd number ______________________

f Spinning an even number ______________________

2 Shade counters in each bag to match the probabilities described.

a One in four chance of picking a white counter

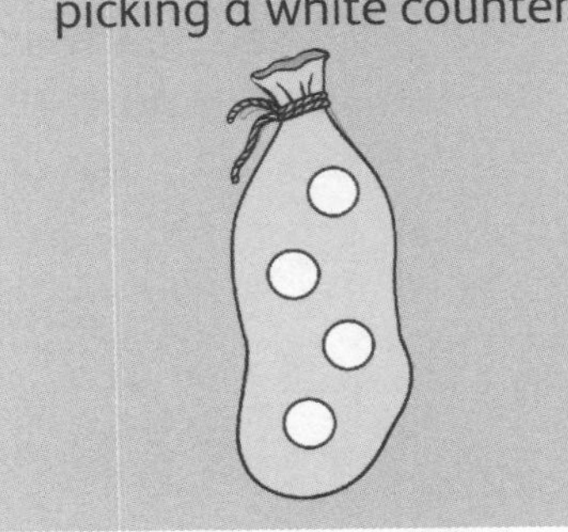

b One in two chance of picking a white counter

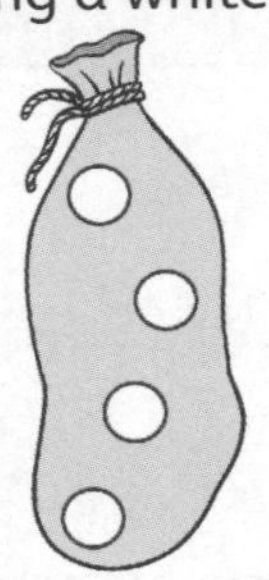

c A 70 % chance of picking a black counter

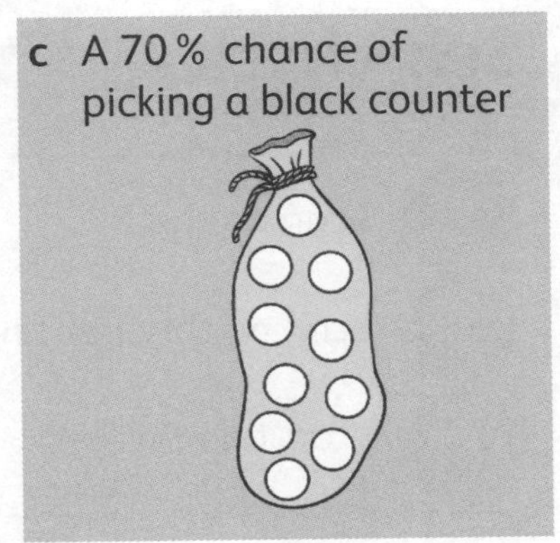

d A 50 % chance of picking a black counter

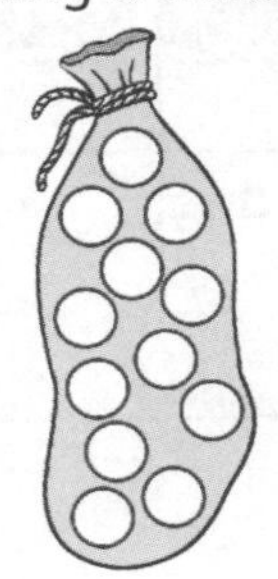

e A greater chance of picking a black than a white counter

f A 40 % chance of picking a white counter

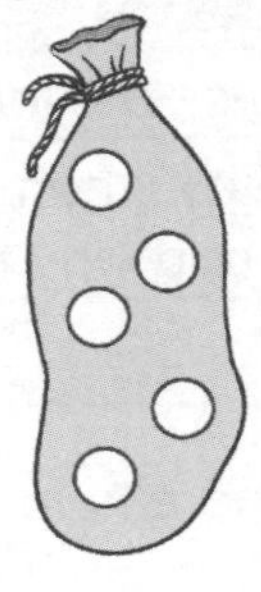

3 In the space below, draw or describe four different mutually exclusive outcomes. Use events from real life or design spinners or other probability experiments.

Unit 6 Probability

Self-check

 I can do this.

 I can do this, but I need to keep trying.

 I can't do this yet.

What can I do?	😉	😐	🙁
1 I can use the language associated with probability and proportion.			
2 I can say when and why two events can happen at the same time and when and why they cannot.			

I need more help with:

____

____

____

Unit 7 Number

Can you remember?

Read these numbers. Then write them using digits.

a Zero point three five

b Thirteen point zero three five

c Three point five zero three

Numbers and place value

1 Fill in the table to show decomposing and regrouping.

Number	14.325	25.609
Decompose as ☐ + ☐ + ☐ …	14.325 =	25.609 =
Regroup as shown	14.325 = 14.3 + ☐ 14.325 → ☐ tenths, ☐ thousandths	25.609 = 10 + 5.6 + ☐ 25.609 → ☐ hundredths, ☐ thousandths

2 **a** Decompose the number –3.252 on this number line.

0

b What is the value of each digit that is repeated?

☐ and ☐

Rounding decimal numbers

1 Round these decimal numbers. Look at the example first.

Number	Rounded to the nearest tenth	Rounded to the nearest whole number	Rounded to the nearest 10
12.39	12.4	12	10
11.11			
19.19			
55.55			
13.99			
29.99			

2

I have 10 cents more than David. My amount of money rounds to $5.

My amount of money rounds to $4.

In the table, show six different amounts that Maris and David could have.

Maris						
David						

Finding common multiples and common factors

1 The array shows that 15 is a common multiple of 3 and 5.

a What are the next three common multiples of 3 and 5?

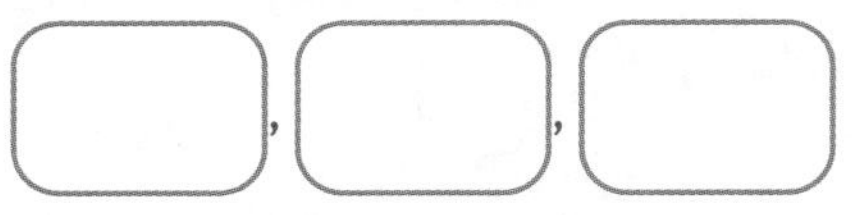

b What **conjectures** or **generalisations** can you make about common multiples of 3 and 5? How might the array help you?

2 **a** Use what you know about the common multiples of 3 and 5 to find the first three common multiples of 2, 3 and 5.

☐, ☐ and ☐

b What do you notice about the common multiples you found in questions **1** and **2a**? Can you explain what you have found?

____

____

____

3 Play this game with a partner. Take turns.
You each need a different colour of pencil.
Choose a square and a circle from below.
Say all the common factors of your pair of numbers.
Your partner must check your answers.
Colour in both shapes if you are correct.
Do this five times each.
The player with the most shapes in their colour wins!

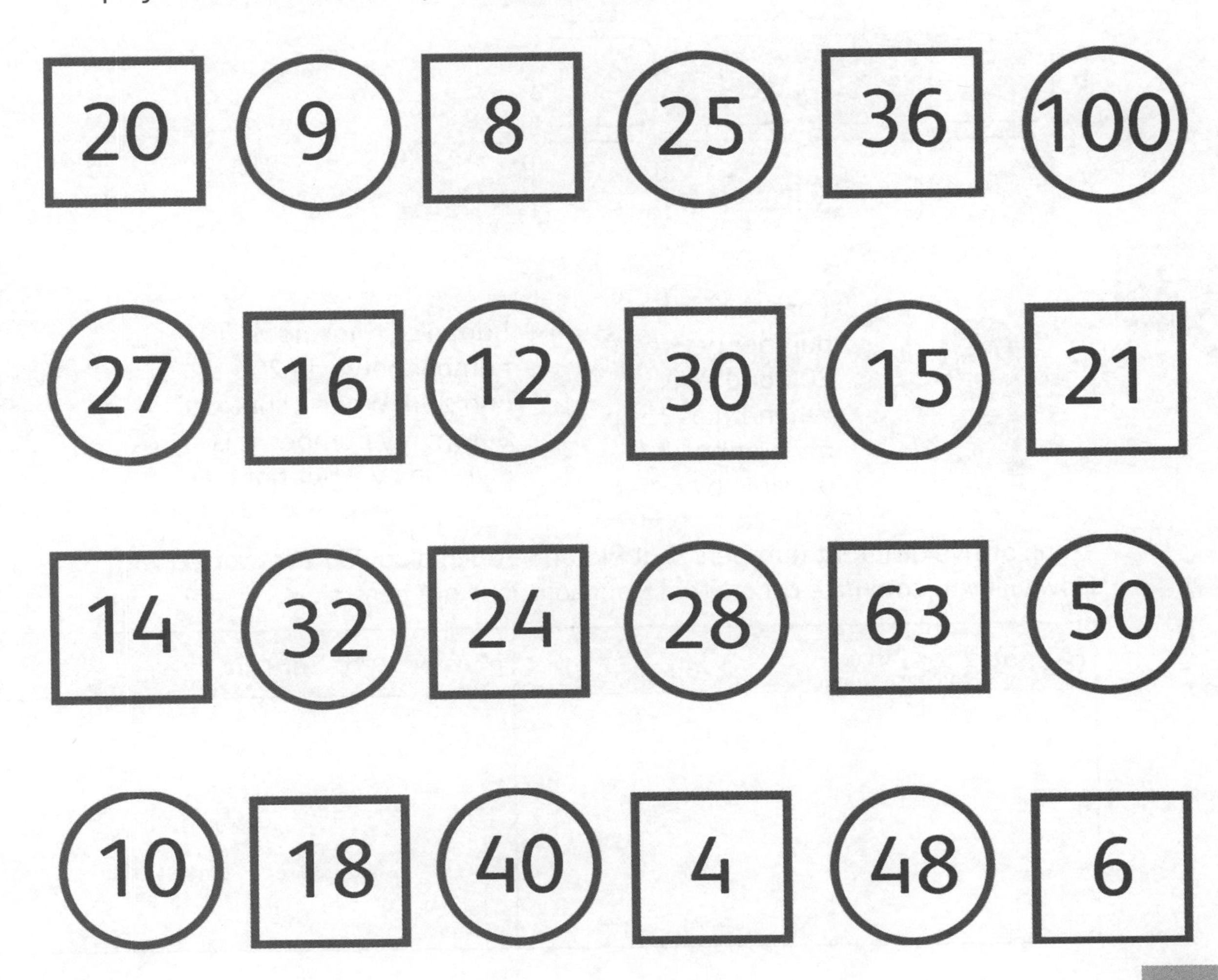

Using tests of divisibility

Use five of the six numbers in the diagram below so that:
- the numbers in the four corners total a number that is divisible by 9
- the three numbers on each diagonal total a number that is divisible by 3 but is not divisible by 9.

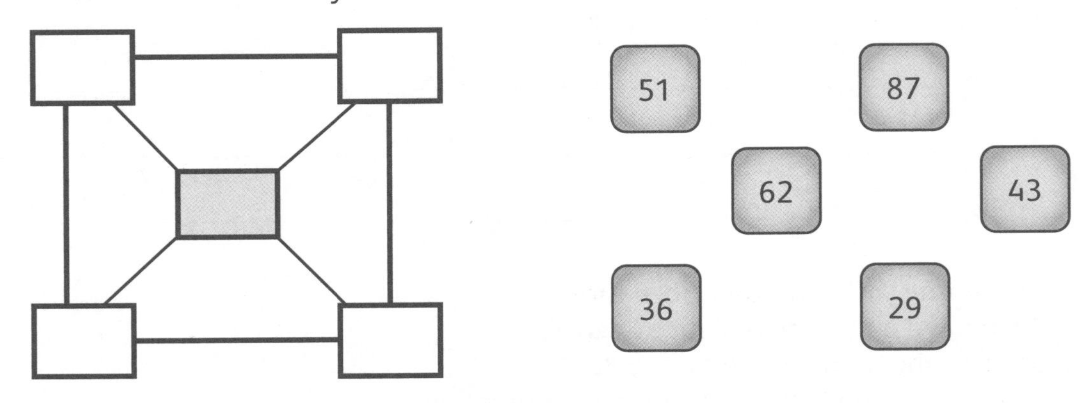

2 Write digits in the boxes to make numbers that are divisible by 9.

I am thinking of a number between 200 and 250. When I add 15 to my number, it is divisible by 6.

I am also thinking of a number between 200 and 250. When I subtract 8 from my number, it is divisible by 3 but not by 6.

Think of five different numbers that Pia and Sanchia could have started with.
How will you **convince** others that your solutions are possible?

Numbers for Pia	Numbers for Sanchia

Unit 7 Number

Self-check

 I can do this.

 I can do this, but I need to keep trying.

 I can't do this yet.

	What can I do?			
1	I can state and explain the value of each digit in decimals (tenths, hundredths and thousandths).			
2	I can compose and decompose numbers using place value, including decimals (tenths, hundredths and thousandths).			
3	I can regroup numbers, including decimals (tenths, hundredths and thousandths) in a variety of ways, for example: 9.234 = 4 + 5.2 + 0.034 or 8.004 + 1.23.			
4	I can multiply and divide whole numbers and decimals by 10, 100 and 1 000.			
5	I can round numbers with two decimal places to the nearest tenth or whole number.			
6	I can explain that a common multiple is a number that is a multiple of two or more numbers.			
7	I can list at least three common multiples of two or more numbers.			
8	I can explain that a common factor is a number that is a factor of two or more numbers.			
9	I can list all the common factors of two or more numbers.			
10	I can say whether or not a number is divisible by 3, 6 or 9.			
11	I can fill in digits to make a number divisible by 3, 6 or 9, for example, put a digit in the box to make a number divisible by 9: 23[3]1.			

I need more help with:

____

____

____

Unit 8 The coordinate grid

Can you remember?

Draw a square with vertices at (2, 2), (2, 9) and (9, 2).

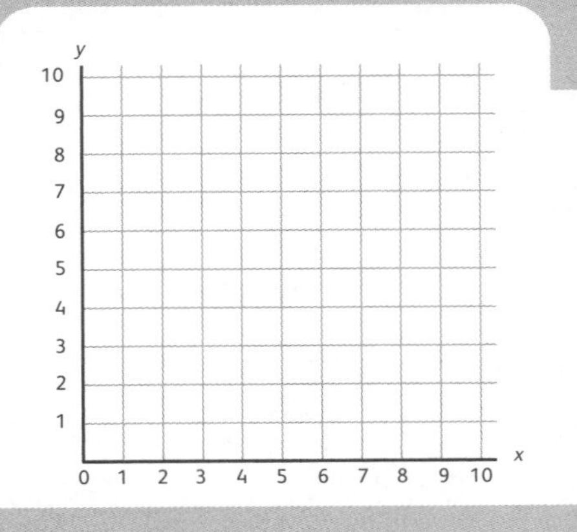

Write the coordinates of the fourth vertex.

Four quadrants

1 Plot the following points on the grid shown below.
Join the points in the order that they are written to make a shape.
Name each shape.

a Shape A: (3, 3), (3, –3), (–3, –3), (–3, 3) Name: ____________________

b Shape B: (–6, 0), (0, 6), (8, 0), (0, –6) Name: ____________________

c Shape C: (0, –10), (–10, 0), (0, 10), (8, 4), (8, –4) Name: ____________________

2 Mark any parallel lines on the shapes using >, >>, >>>, >>>>, >>>>> or >>>>>>.

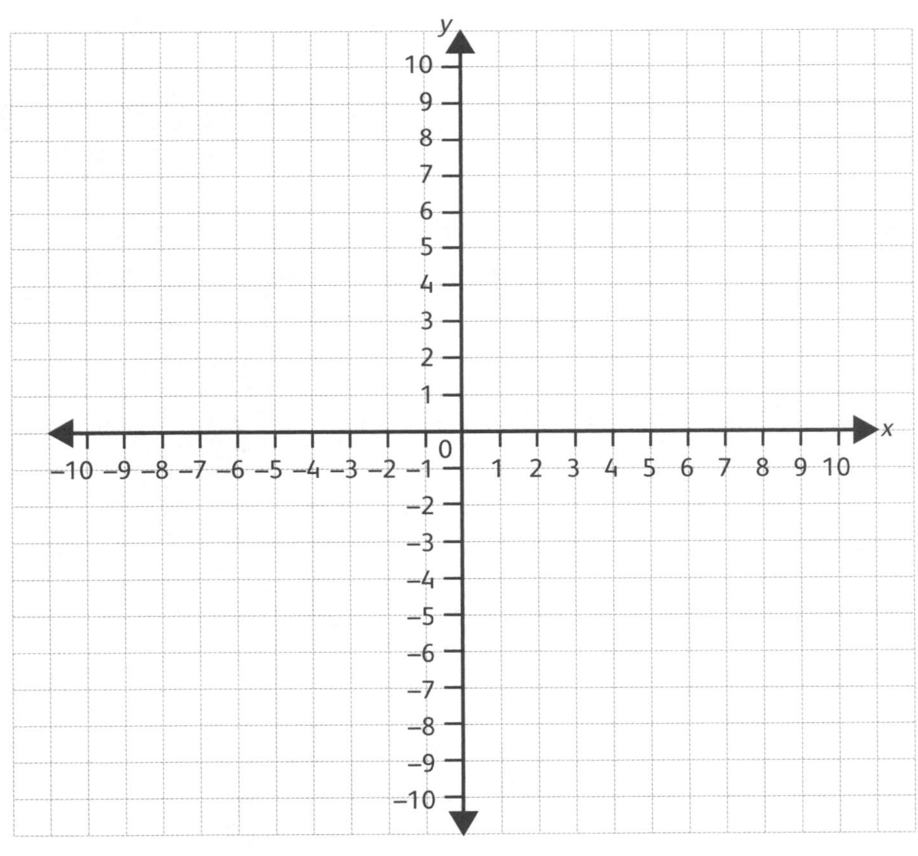

Translations on a coordinate grid

1 Look at the grid and answer the questions below.

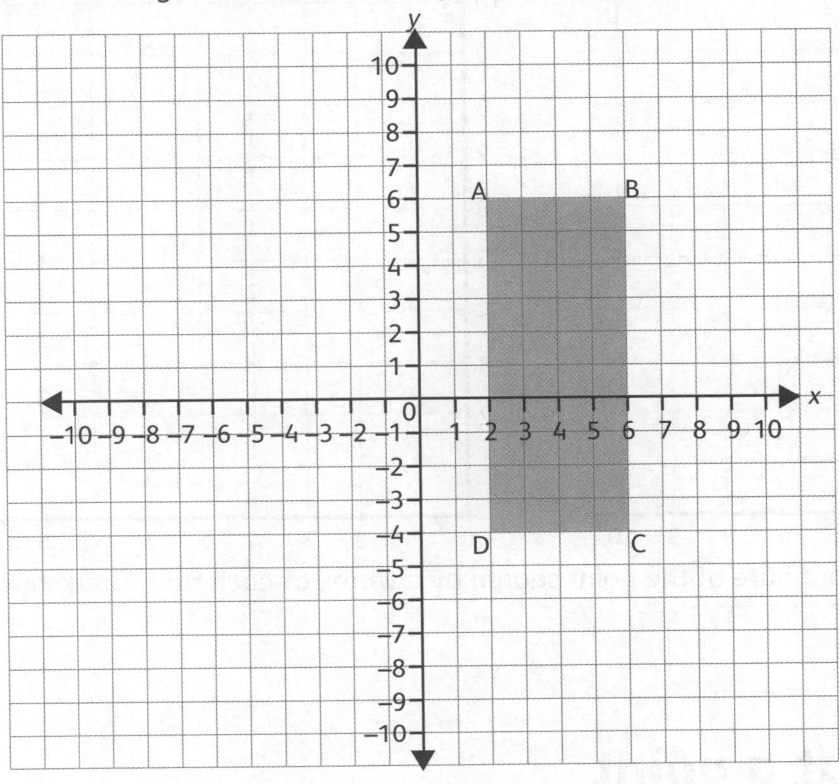

a Write the coordinates of the vertices at points A, B, C and D.

A = (________, ________) B = (________, ________)

C = (________, ________) D = (________, ________)

b The rectangle is translated five squares left and three squares up.
Predict what the coordinates will be for each new vertex.

A = (________, ________) B = (________, ________)

C = (________, ________) D = (________, ________)

c Draw the new shape on the grid.
Write the actual coordinates of each new vertex.

A = (________, ________) B = (________, ________)

C = (________, ________) D = (________, ________)

d Compare these coordinates with your predictions.
Describe the method you used to make the prediction.

____

____

____

2 a Translate the rectangle three squares right and two squares down.
Translate the triangle four squares left and three squares up.

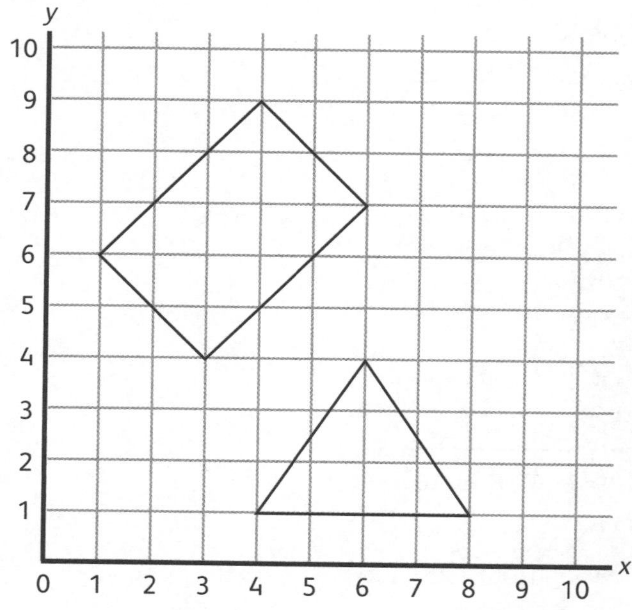
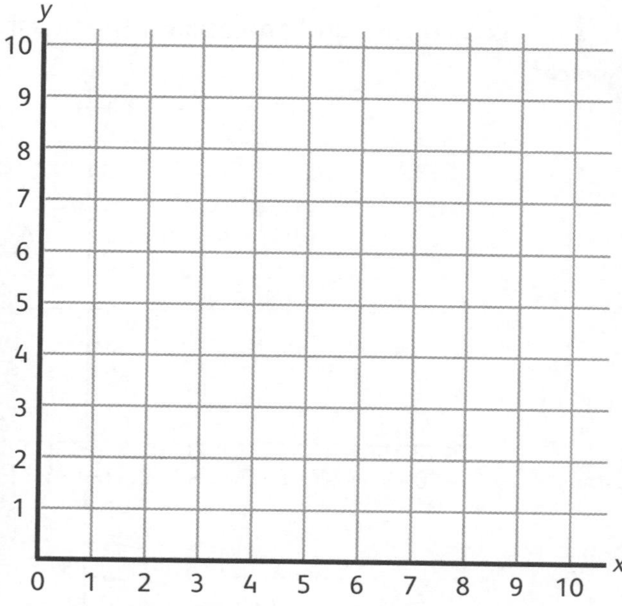

b What is the coordinate of the point shared by a vertex of each translated shape?

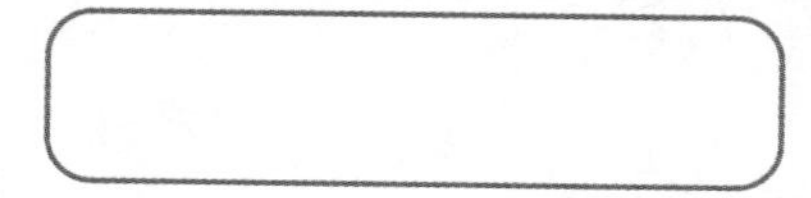

Rotation about a point

1 Rotate each shape by 90 degrees **clockwise** around the vertex marked with a dot.

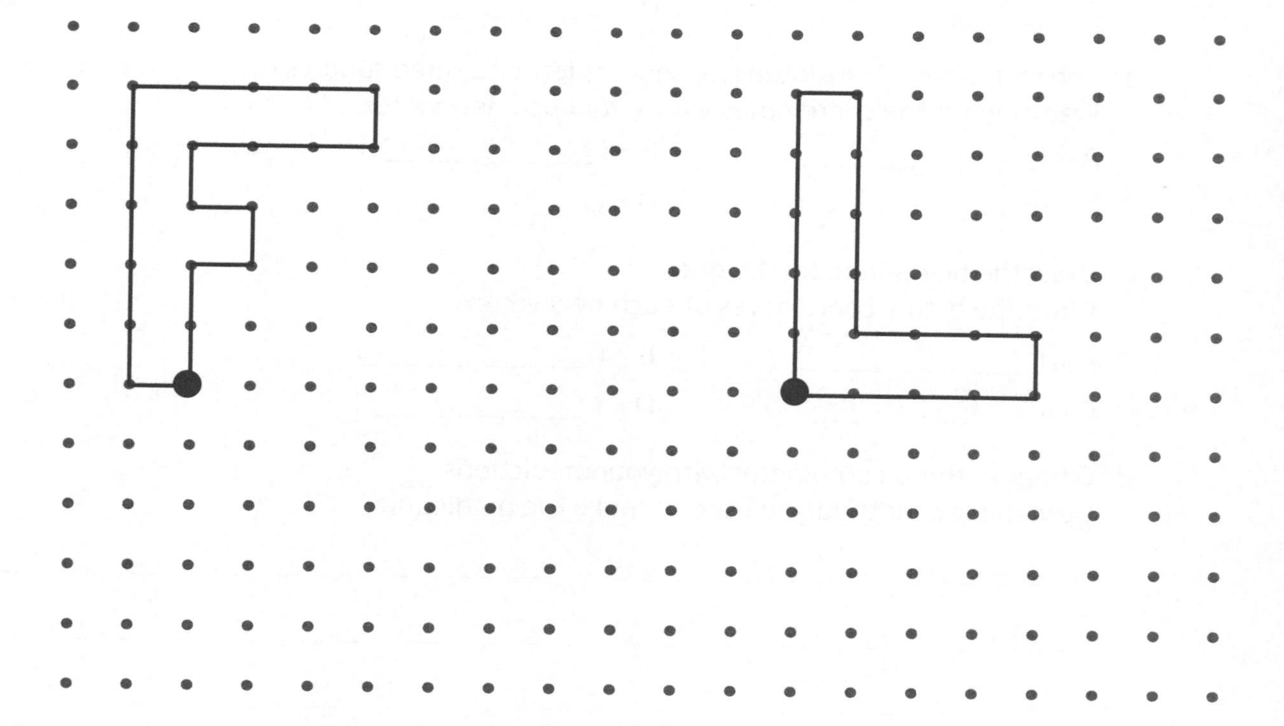

2 Rotate each shape by 90 degrees **anticlockwise** around the vertex marked with a dot.

3 Create your own letter designs and rotate the shape through 90 degrees.
Repeat each rotation four times.

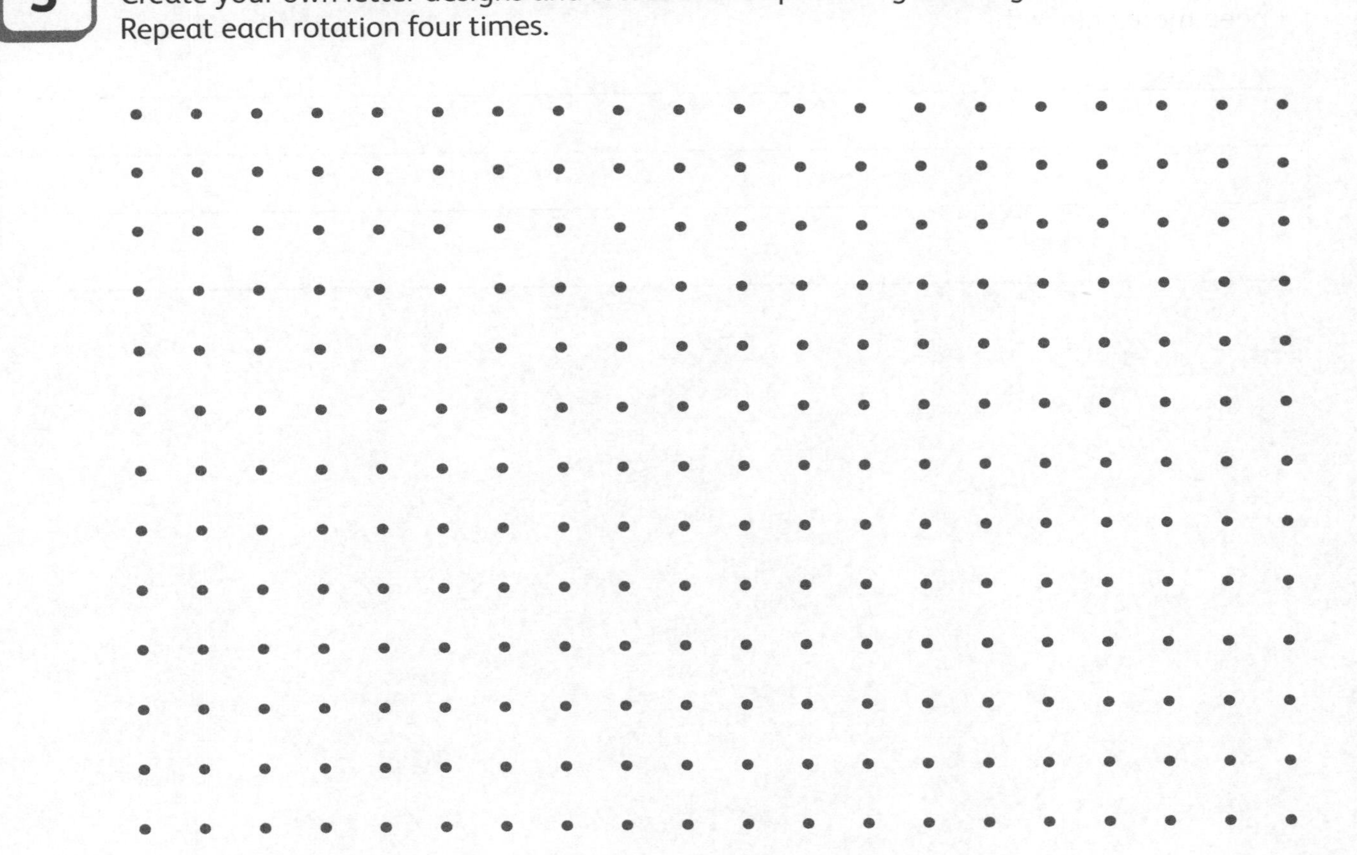

Unit 8 The coordinate grid

Self-check

 I can do this.

 I can do this, but I need to keep trying.

 I can't do this yet.

What can I do?			
1 I can read coordinates in all four quadrants (with the help of a grid).			
2 I can plot points to form lines and shapes in all four quadrants by drawing on my knowledge of 2D shapes and coordinates.			
3 I can translate 2D shapes on coordinate grids.			
4 I can rotate shapes 90° around a vertex (clockwise and anticlockwise).			

I need more help with:

____

____

____

Unit 9 Calculation

Can you remember?

Which of these divisions will leave a remainder? Use tests of divisibility to help you.

	TICK (✓) Yes	TICK (✓) No		TICK (✓) Yes	TICK (✓) No
a 486 ÷ 5			**b** 328 ÷ 4		
c 1000 ÷ 25			**d** 815 ÷ 3		
e 998 ÷ 2			**f** 1 489 ÷ 9		

Using addition and subtraction

1

a Use letters to represent the perimeter of the square as an addition.

___

b Now record the perimeter as a multiplication.

___

c Calculate the perimeter of each square when:

i s = 8 metres	**ii** s = 12 metres	**iii** s = 25 metres

Look at this graph of temperature readings. The temperature difference between Tuesday and Wednesday is the same as the temperature difference between Wednesday and Thursday. On both pairs of days, the temperature changed by 3°.

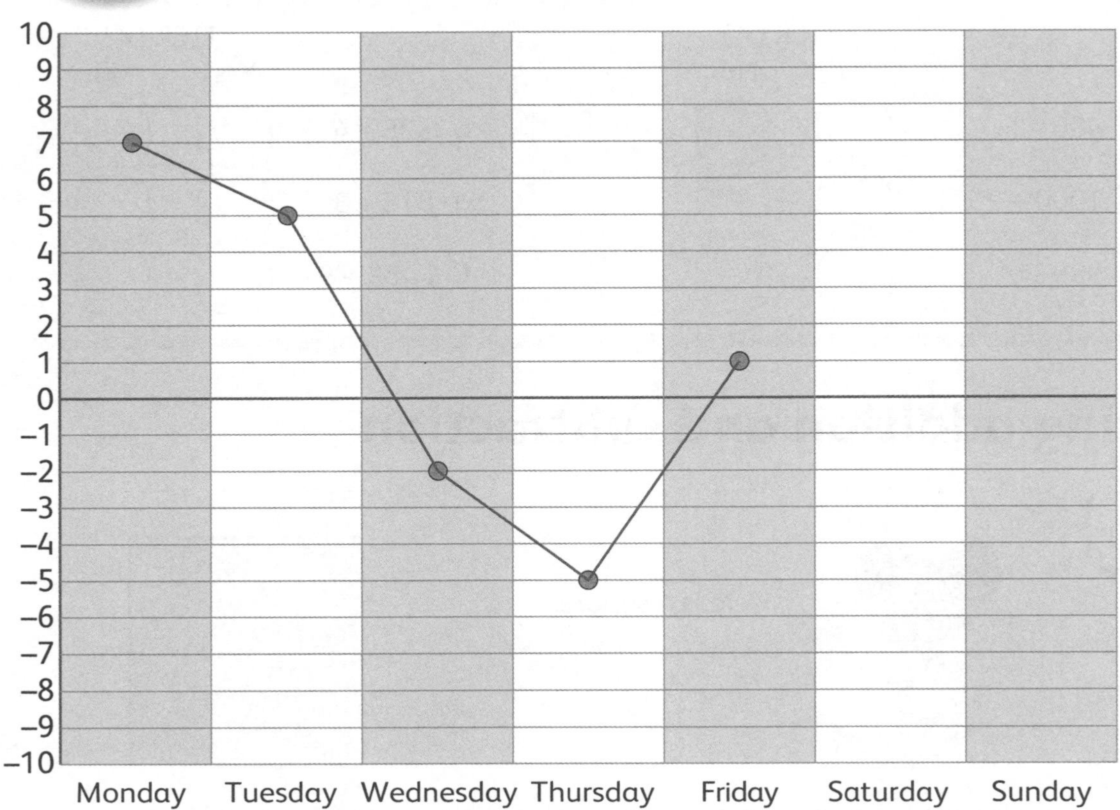

Do you agree with Jin? Explain your answer.

____

____

Adding and subtracting decimal numbers

1 Do the calculation in **bold** first. This will help you to do the other three.

a

77 + 23 = ☐	7.7 + 2.3 = ☐
0.77 + 0.23 = ☐	7.7 + ☐ = 10

b

36 + 64 = ☐	3.6 + 6.4 = ☐
0.36 + 0.64 = ☐	0.036 + 0.064 = ☐

c

375 − 152 = ☐	37.5 − 15.2 = ☐
3.75 − 1.52 = ☐	0.375 − 0.152 = ☐

2 Mix up a set of 0 to 9 digit cards. Place them face down on the table. Take the first five cards from the pile.

a Arrange the cards in each addition calculation to make the smallest and largest totals.

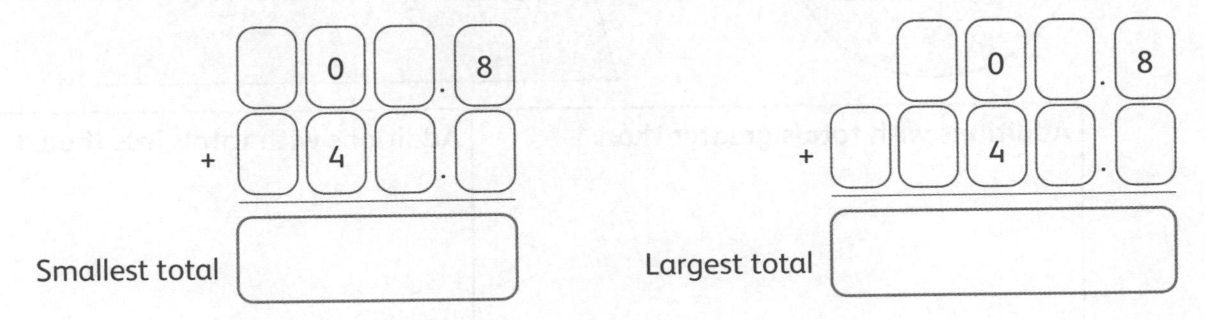

Smallest total ☐ Largest total ☐

b Arrange the cards in these subtraction calculations to make the smallest and largest differences.

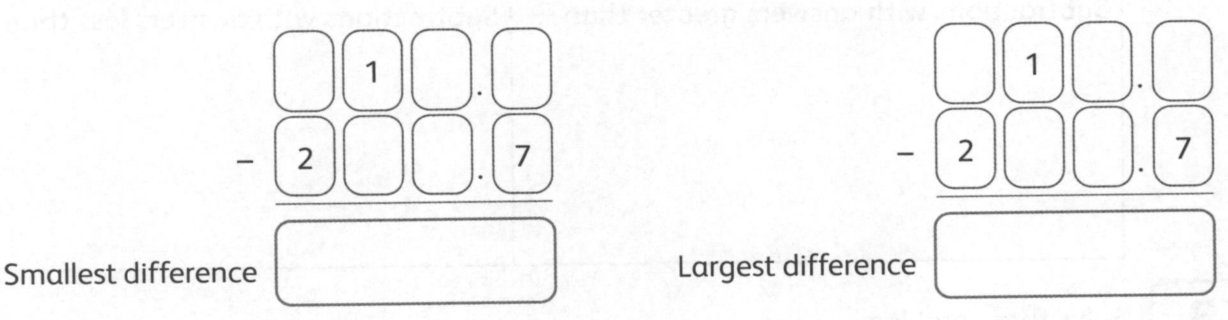

Smallest difference ☐ Largest difference ☐

Adding and subtracting fractions

1 **a** Use the bar models to represent the addition $\frac{2}{3} + \frac{3}{4}$.

Divide the bottom bars to show equivalent fractions with a common denominator.

$\frac{1}{3}$	$\frac{1}{3}$	$\frac{1}{3}$

$\frac{1}{4}$	$\frac{1}{4}$	$\frac{1}{4}$	$\frac{1}{4}$

b Complete the calculation: $\frac{2}{3} + \frac{3}{4} =$ ☐ $\frac{☐}{☐}$

Choose a fraction from each shape to make additions and subtractions.
Use estimates to decide which box to put them in.
Write at least two calculations in each box. Solve each one to check.

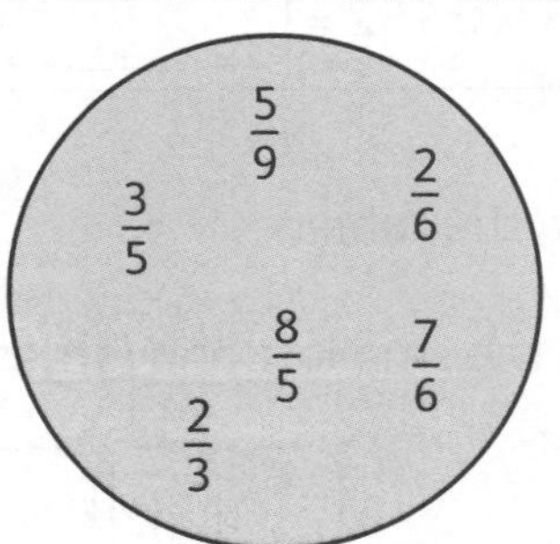

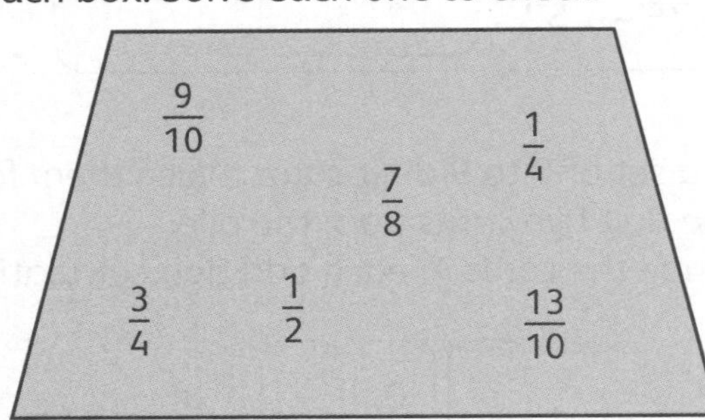

Additions with totals greater than 1	Additions with totals less than 1
Subtractions with answers greater than $\frac{1}{2}$	**Subtractions with answers less than $\frac{1}{2}$**

3 Solve these problems.

a A recipe requires $\frac{1}{2}$ kg of potatoes and $\frac{2}{5}$ kg of carrots. Their total mass is: ________ kg

b A rope is equivalent in length to $\frac{9}{10}$ m. A ribbon is $\frac{3}{5}$ m shorter than the rope.

How long is the ribbon? The ribbon is ________ metres long.

Multiplying decimal numbers

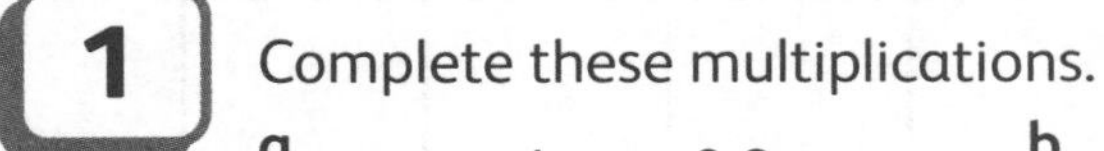

Complete these multiplications.

a

×	4	0.8
6		

Product: ______________

b

×	4	0.8
20		
6		

Product: ______________

c

×	7	0.4	0.08
20			
6			

Product: ______________

2 The pictogram shows the population of different cities.

Population of different cities

City A ●●●●●●●●●
City B ●●●●●
City C ●●●●●●●●
City D ●●●●●●●

What is the population of each city when ● has these different values?

Population	City A	City B	City C	City D
When ● has the value of 0.5 million				
When ● has the value of 0.25 million				
When ● has the value 1.5 million				

Dividing whole numbers up to 1000

1 **a** Circle the divisions with answers of less than 100. Use estimates to help you.

705 ÷ 8 636 ÷ 5 459 ÷ 9 381 ÷ 3 581 ÷ 7

b Calculate each answer to see if you are correct.

★ **c** What **generalisations** can you make about the divisions that have answers of less than 100?

____

____

★ **2** Here are six 1-digit numbers.

3 4 5 7 8 9

Choose four digits. Arrange them in the division calculation so that your answer has a remainder. Turn the remainder into a fraction. What is it?

☐) ☐ ☐ ☐

____

3 Decide whether you need to round the answer or turn the remainder into a fraction.

a A box of eight games costs $108. How much does one game cost? ______________

b A bus can carry 25 people. How many buses do 240 people need? ______________

c A farmer packs eggs into boxes of six. She has 195 eggs.
How many full boxes of eggs can the farmer pack? ______________

Using brackets

1 a Write the answers to these calculations.
Remember to look at the brackets first. Work down each column.

$9 \times (5 \times 6) =$ ☐	$9 \times (5 + 6) =$ ☐
$(9 \times 5) \times 6 =$ ☐	$(9 \times 5) + 6 =$ ☐
$(6 \times 9) \times 5 =$ ☐	$(6 + 9) \times 5 =$ ☐
$6 \times 9 \times 5 =$ ☐	$6 \times (9 + 5) =$ ☐

b

That's strange. I think all the answers in the first column are the same, but not in the second column! Why could this be?

How can you explain this to Sanchia?

____

____

____

____

2 Use brackets to help you keep track of your calculations in these problems.

	Problem	My calculation
a	Pencils come in packs of 25. Guss buys five packs. He gives Elok 10 pencils from each pack. How many pencils does Guss have left?	☐ pencils
b	A café sells cupcakes. Each cupcake costs 35 cents to make. They sell nine cupcakes at 85 cents each. How much profit do they make in total?	☐ cents

Unit 9 Calculation

Self-check

 I can do this.

 I can do this, but I need to keep trying.

 I can't do this yet.

What can I do?			
1 I can estimate, add and subtract integers, including finding differences between positive and negative integers, and two negative integers.			
2 I can use letters to represent quantities that vary in addition and subtraction calculations.			
3 I can estimate the sums or differences between numbers with the same or different numbers of decimal places.			
4 I can add and subtract numbers with the same or different numbers of decimal places.			
5 I can estimate the sums and differences between fractions with different denominators.			
6 I can add and subtract fractions with different denominators.			
7 I can estimate the product of a number with one or two decimal places with a 1-digit or 2-digit whole number.			
8 I can multiply numbers with one or two decimal places by 1-digit or 2-digit whole numbers.			
9 I can apply the laws of arithmetic to simplify calculations.			
10 I can estimate the result of dividing whole numbers up to 1 000 by 1-digit or 2-digit whole numbers.			
11 I can divide whole numbers up to 1 000 by 1- or 2-digit whole numbers.			
12 I can explain the difference between giving an answer with a remainder or turning the remainder into a fraction (when I can see that it is no longer a remainder), for example: $36 \div 8 = 4$ r 4 or $4\frac{1}{2}$.			
13 I can explain how using brackets can change the order of operations.			

I need more help with:

___

___

___

Unit 10 Probability

Can you remember?

Look at the bag of cubes and write **true** or **false** for each statement.

a There is a two in four chance of picking a white cube. ________

b There is a 75 % chance of picking a black cube. ________

c There is less than a 25 % chance of picking a grey cube. ________

d The chance of picking a white cube is equal to the chance of picking a grey cube. ________

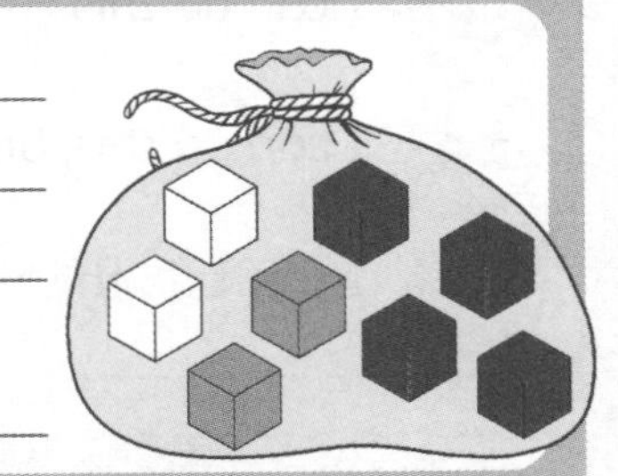

Probability experiments

1 Play a spinning game. Spin both spinners below. Then add the score. Keep a tally of your scores.

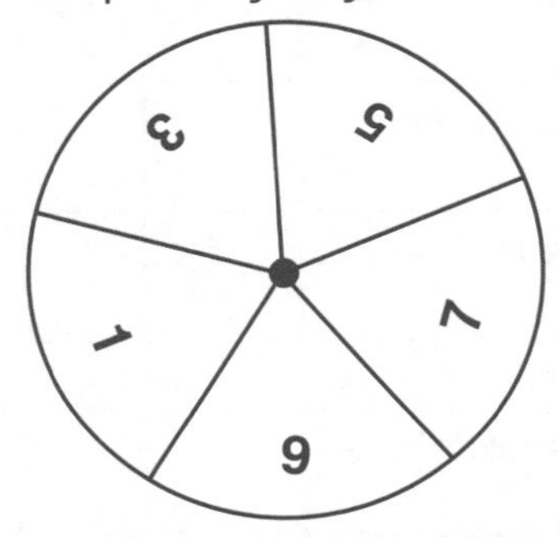

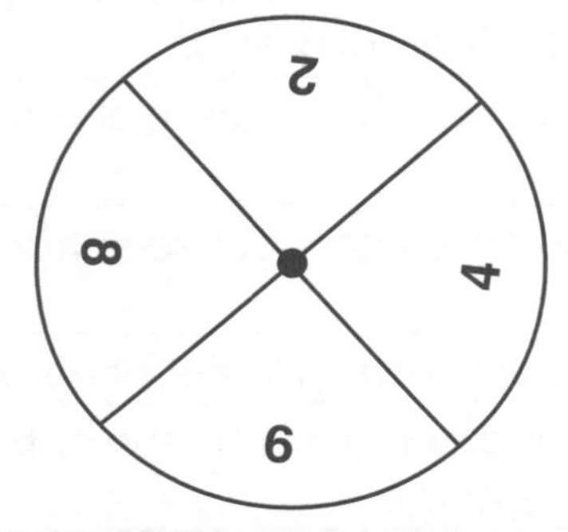

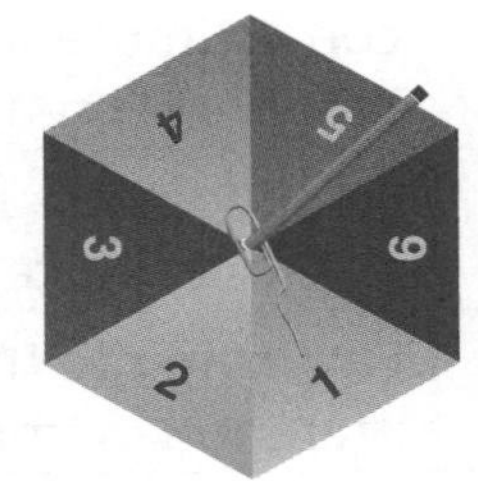

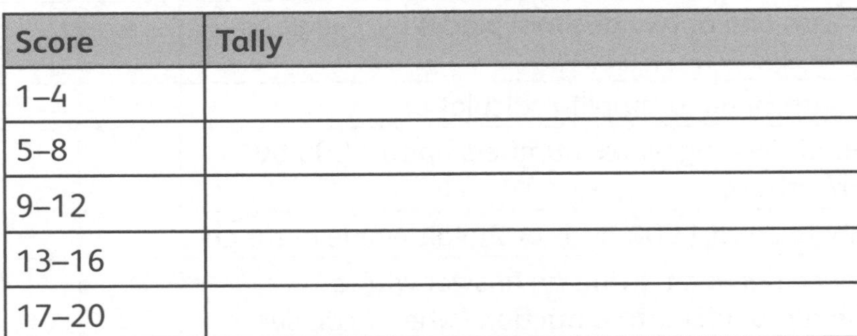

Score	Tally
1–4	
5–8	
9–12	
13–16	
17–20	

a Complete this sentence:

I predict that the most common score will be between ☐ and ☐

because ______________________________.

b Spin 10 times. Was your prediction correct? ________________

c Now spin 20 times more.
Which is the most common range of scores?

Complete these two spinners by writing a number in each section.
Decide on a rule for playing the game, for example:

- Add the two scores.
- Subtract the two scores.
- Multiply the two scores.

My rule is:

____

____

Record your results in this table.

Score range	Frequency

Use your **conjecturing** skills to predict the outcomes you expect.

I predict that: ________________________________

____

____

____

Complete the experiment 50 times. Tally your scores into the table above.
Compare the results with your predictions. Explain any differences.

____

____

____

____

____

Unit 10 Probability

Self-check

 I can do this.

 I can do this, but I need to keep trying.

 I can't do this yet.

What can I do?	I can do this	I need to keep trying	I can't do this yet
1 I can say why a small number of trials may not give a clear indication (outcome or understanding) of the real probability.			
2 I can carry out and record the results of chance experiments or simulations (models), using small and large numbers of trials.			
3 I can predict, analyse and describe the frequency of outcomes using the language of probability.			

I need more help with:

____

____

____

Unit 11 Fractions, decimals, percentages, ratio and proportion

Can you remember?

Write each set of fractions in their simplest form.

a	b	c
$\frac{3}{12}$ =	$\frac{2}{10}$ =	$\frac{25}{100}$ =
$\frac{4}{12}$ =	$\frac{4}{10}$ =	$\frac{75}{100}$ =
$\frac{6}{12}$ =	$\frac{6}{10}$ =	$\frac{80}{100}$ =
$\frac{9}{12}$ =	$\frac{5}{10}$ =	$\frac{150}{100}$ =

More about fractions as operators

1 Which is the larger quantity each time? Tick (✓) the correct box.

a $\frac{3}{4}$ of \$64 ☐ or $\frac{4}{3}$ of \$39 ☐

b $\frac{3}{5}$ of 70 cm ☐ or $\frac{5}{3}$ of 45 cm ☐

c $\frac{9}{10}$ of 40 kg ☐ or $\frac{10}{7}$ of 21 kg ☐

d $\frac{5}{8}$ of 72 litres ☐ or $\frac{8}{5}$ of 25 litres ☐

2 Elok measures her pencil case. It is 25 cm long.
David measures the bookcase. It is $\frac{12}{5}$ of the length of the pencil case.
Pia measures her pen. It is $\frac{3}{5}$ of the length of the pencil case.
What are the lengths of the bookcase and the pen?

Bookcase: ________ cm Pen: ________ cm

3 How much heavier is $\frac{4}{9}$ of 63 kg than $\frac{5}{2}$ of 10 kg? ________ kg

Comparing and ordering decimal numbers

1 Begin at **Start**. End at **Finish**. Make a path of numbers through touching circles, so that they are in order from smallest to largest. Colour them in to show the path.

a **Start**

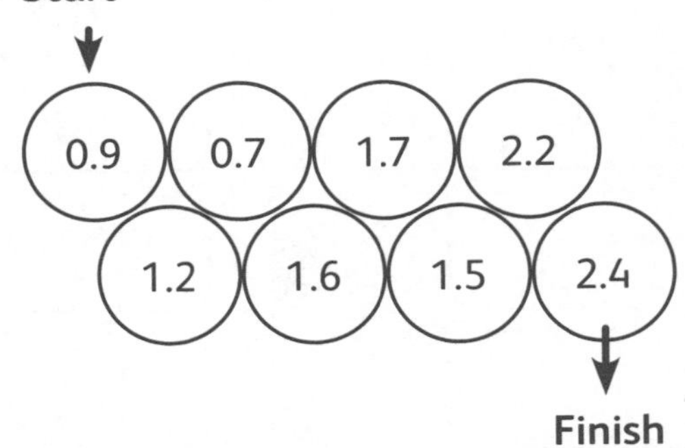

Finish

b **Start**

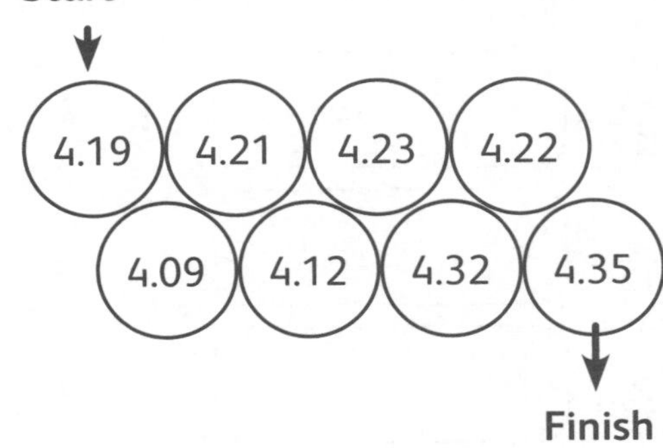

Finish

2 Use the symbol **>** or **<** to make each statement true.

a 3.43 ☐ 3.45

b 34.5 ☐ 34.3

c 51.98 ☐ 51.89

d 69.6 ☐ 69.7

e 5.4 ☐ 32.3

3 The table shows the length of throws in a javelin competition.

Athlete	A	B	C	D	E
Length of throw	64.58 m	65.84 m	58.46 m	68.45 m	58.64 m

Order the throws from longest to shortest. Write only the letters of the athletes.

____

Longest **Shortest**

Using decimals and mixed units for time

1 Draw lines to match equivalent units of time.

3.5 hours	3 hours and 15 minutes
3.75 hours	3 hours and 36 minutes
3.25 hours	3 hours and 45 minutes
3.2 hours	3 hours and 30 minutes
3.6 hours	3 hours and 12 minutes

2 I think 1.4 hours is the same as 1 hour and 40 minutes.

Draw your own bar models to explain Sanchia's mistake.

3 A car takes 9 minutes to travel 4 km. How long does it take for the car to travel 1 km? Give your answer in minutes and in mixed units.

________ minutes ________ minutes and ________ seconds

Direct proportion

1

Strawberry ice cream

300 g strawberries	________
90 g sugar	________
250 ml milk	________
250 ml cream	________
1 vanilla pod	________
10 egg yolks	________
2 tablespoons lemon juice	________
Serves: 6	Serves: 18

a Rewrite the recipe to serve 18 people. Use the lines provided.

b If we use only 30 g of sugar, how many people will it serve?

____

2 A model boat is nine times as small as this real boat.
What are the dimensions of the model boat?

Height: ______________________

Length: ______________________

Width: ______________________

Equivalent ratios

1

This tile has a pattern of squares and circles.

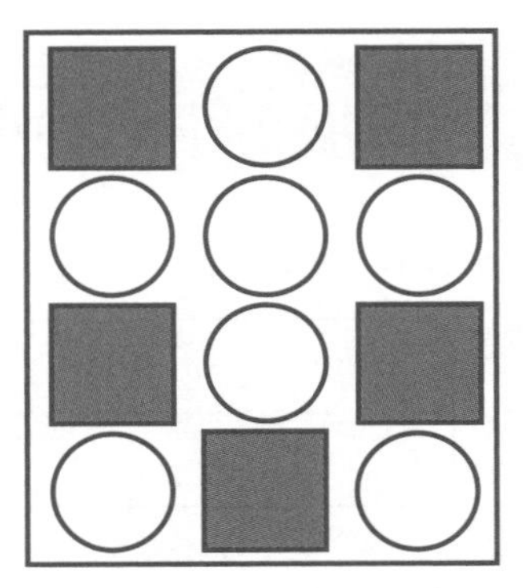

a Complete the sentence:
For every ______ squares, there are ______ circles.

b Now complete the table to show the number of squares and circles each time.

Tiles	1	2	4	8	10	20
Circles	7			56		
Squares		10			50	
Total shapes	12		48			

2 Solve this problem. Jin packs a box of toys in the following ratio:

a How many dolls will there be when there are 30 teddies in the boxes? ______________

b How many teddies will there be when there are 18 lorries in the boxes? ______________

3 The scale on a map is 2 cm : 5 km.
If two places are shown 6 cm apart on the map, how far apart are they in real life?

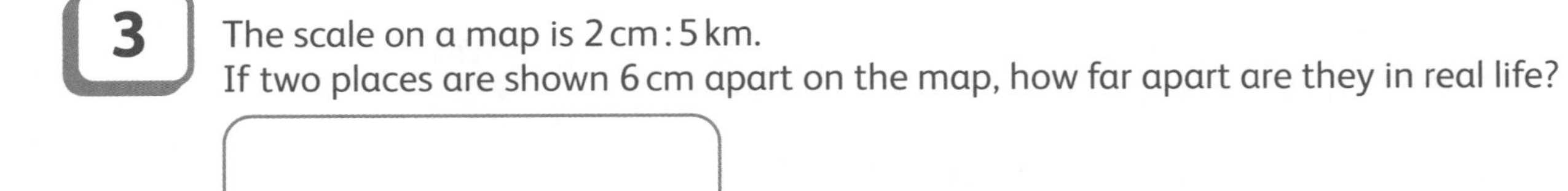

Unit 11 Fractions, decimals, percentages, ratio and proportion

Self-check

 I can do this.

 I can do this, but I need to keep trying.

 I can't do this yet.

See how much you know!

What can I do?			
1 I can solve context problems involving proper and improper fractions as operators.			
2 I can solve problems involving proper and improper fractions as operators.			
3 I can compare and order numbers with one or two decimal places, using the symbols =, > and <.			
4 I can convert between time intervals expressed as decimals and in mixed units.			
5 I can give values of measures that are in proportion, with simple numbers and in context.			
6 I can solve context problems involving simple ratios, for example: A builder uses sand and cement. He uses three bags of sand for every two bags of cement. If there are ten bags of cement, how many bags are sand? (15 bags are sand.)			

I need more help with:

Unit 12 2D and 3D shapes

Can you remember?

Write the name of each 3D shape.

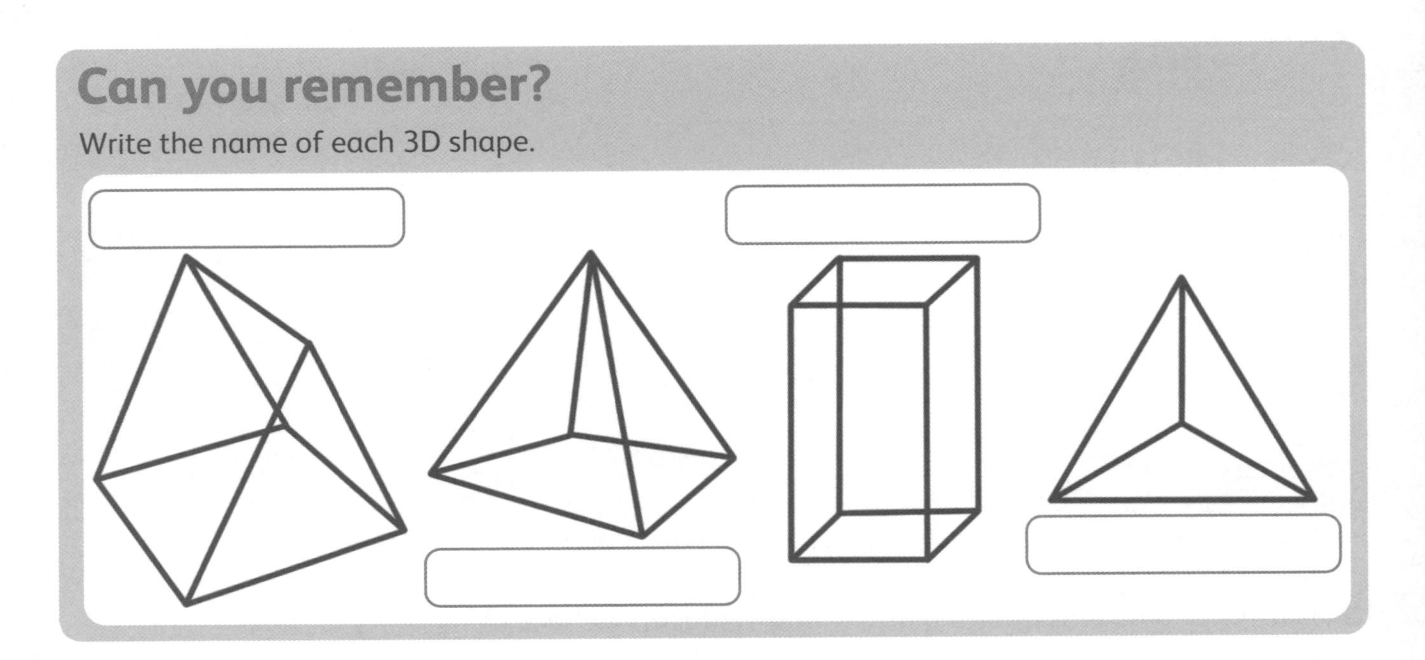

Compound shapes

1 Copy each cuboid made from individual cubes. Use the dot grid.

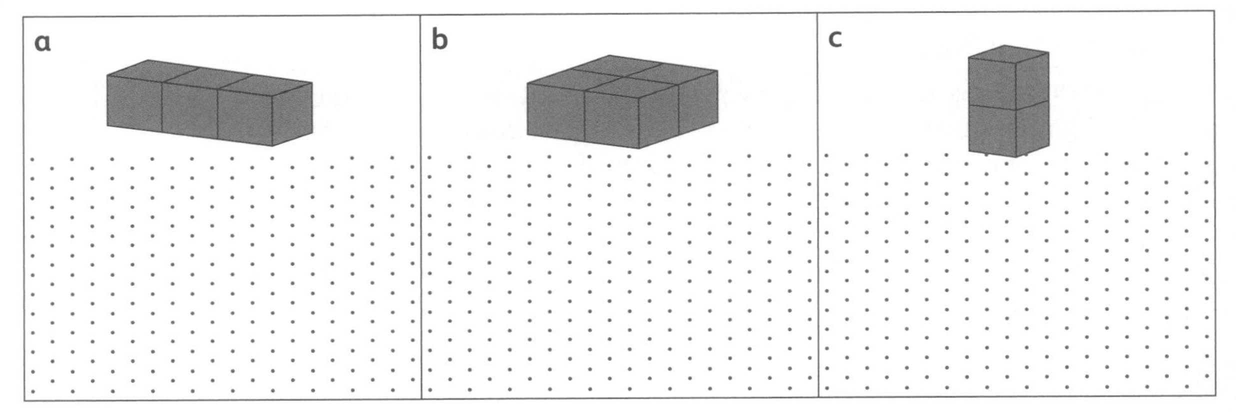

2 Copy each compound shape. Use the dot grid.

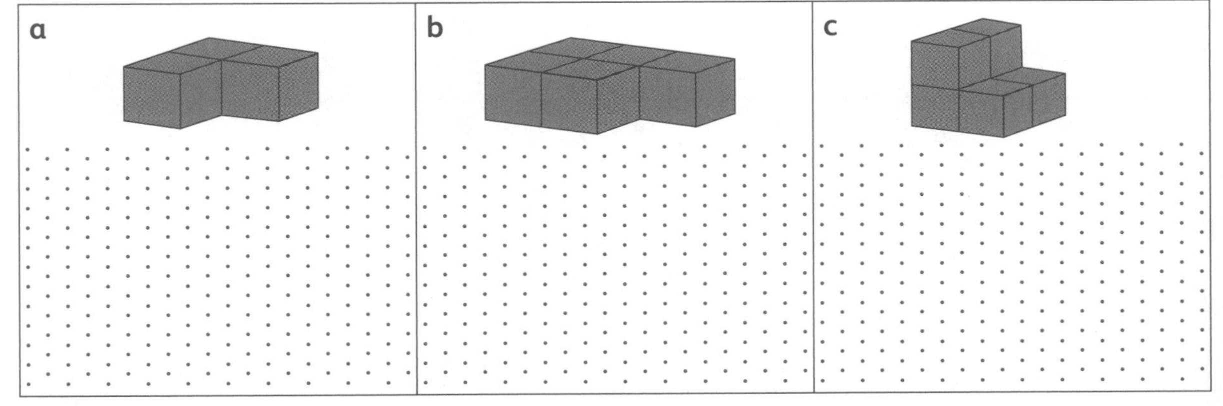

3 Design a cuboid using six cubes. Sketch it below.
Then draw your cuboid as you would see it from two other positions.

4 Design a shape using seven cubes. Sketch it below.
Draw it as you would see it from two other positions.

Nets

1 Look at the nets below and picture each cube in your mind.
Colour in each net so that the faces that share an edge are different colours.
Try to use as few colours as possible.

a

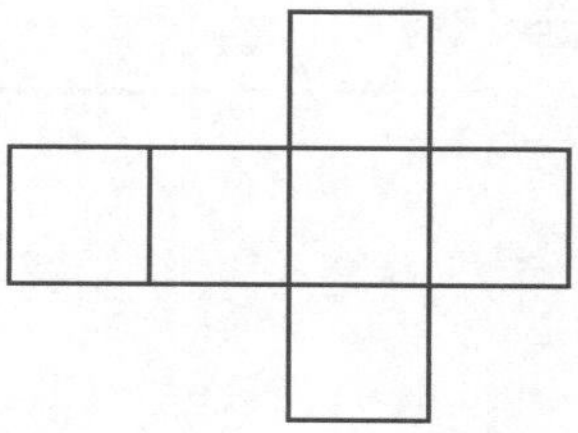

b

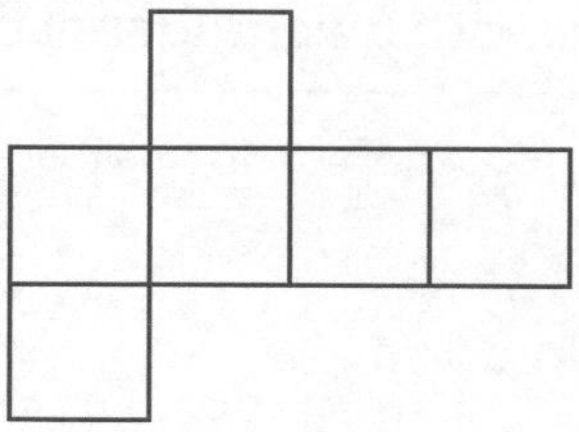

c

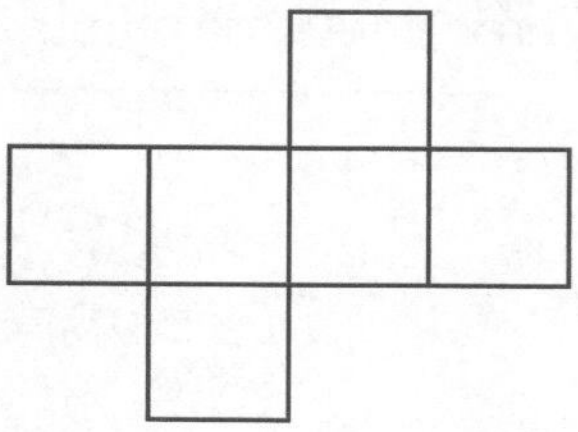

d

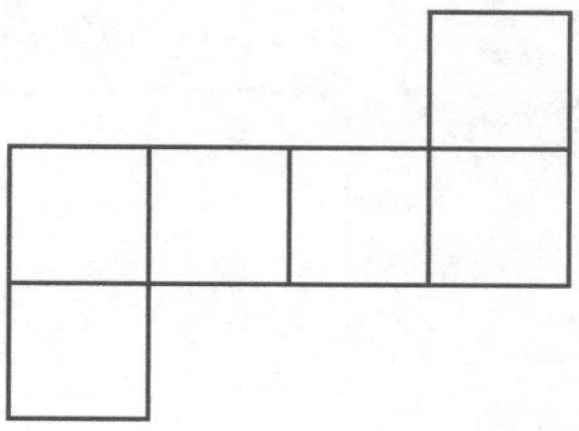

2 Write the name of the 3D shape made by each net.
Colour in each net so that any faces that share an edge are different colours.
Try to use as few colours as possible.

a

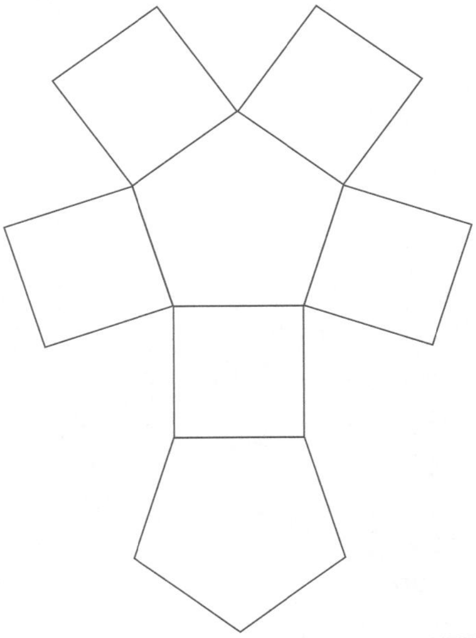

b

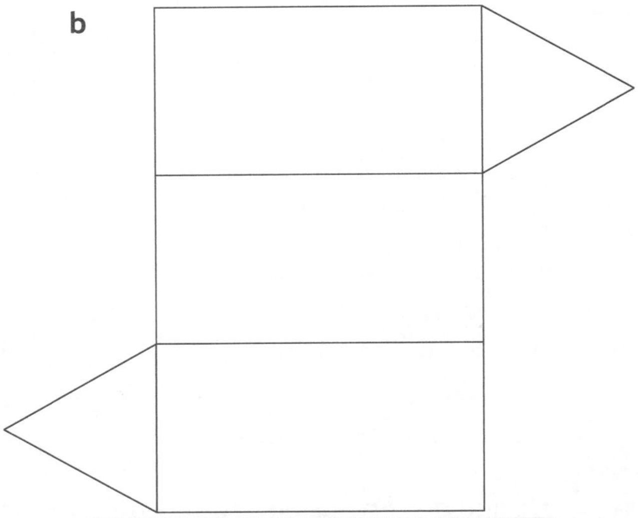

c

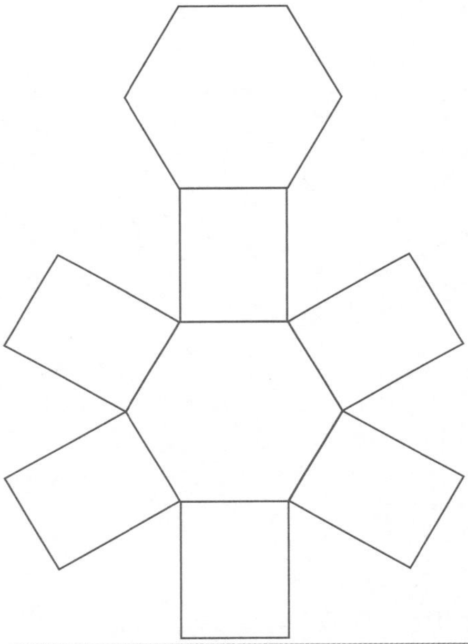

d

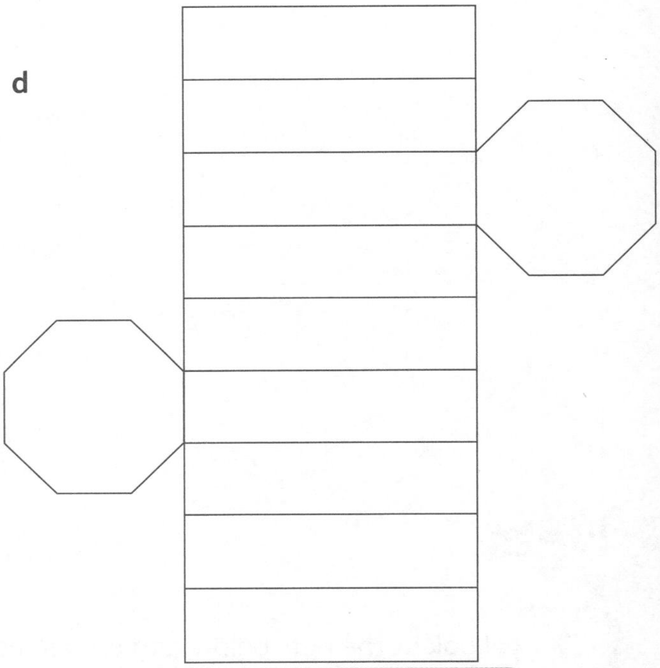

3 Sketch two different nets for pyramids with a pentagonal base.

Unit 12 2D and 3D shapes

Self-check

 I can do this.

 I can do this, but I need to keep trying.

 I can't do this yet.

What can I do?	😊	😐	☹
1 I can identify and describe compound 3D shapes, including isometric representations (which show equal measurements after rotating).			
2 I can sketch compound 3D shapes, including isometric representations (which show equal measurements after rotating).			
3 I can explain the difference between capacity and volume.			
4 I can identify different nets for cubes, cuboids, prisms and pyramids.			
5 I can sketch different nets for cubes, cuboids, prisms and pyramids.			

I need more help with:

____

____

____

Unit 13 Number

Can you remember?

Fill in the missing numbers.

a $0.04 \times \square = 40$	b $\square \div 100 = 5.301$	c $960 \div \square = 0.96$
$400 \div 1\,000 = \square$	$5.301 \times \square = 5\,301$	$\square \times 100 = 9.6$

More about numbers and place value

1 Fill in this table.

	Number	How many tenths?	How many hundredths?	How many thousandths?
a	5.291			
b	42.36			
c	9.4			
d	4.405			

2 Write the value of the underlined digit/s each time.

a 702.135 __________

b 75.67 __________ and __________

c 940.52 __________

d 0.696 __________ and __________

3

I am thinking of a number with three decimal places. One digit has the value 0.06. My number has 45 tenths. Two digits are repeated.

What could Banko's number be? Find all the possible solutions.

Rounding decimal numbers

1 Jin has rounded each mass below to the nearest $\frac{1}{10}$ of a kilogram.

a **Critique** Jin's work: Tick (✓) if correct, or make **improvements**.

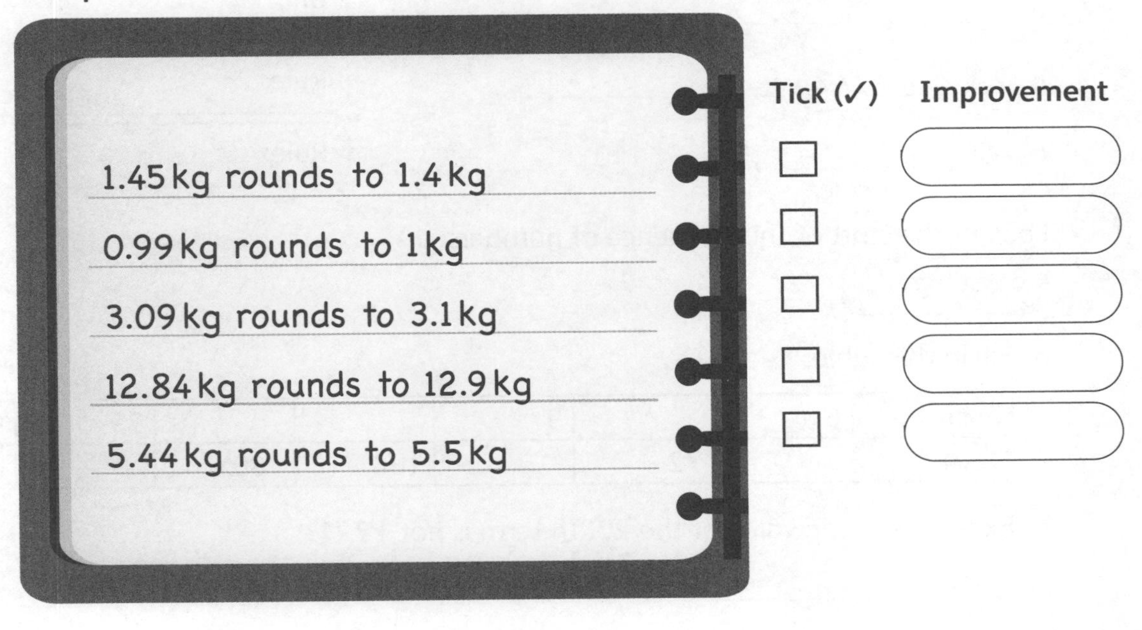

b Make up two more examples of masses like those shown above.

____

____

2

How many possible starting numbers are there?

____

3 David and Maris each have $35. Use rounding to estimate which pairs of tickets show items they can buy. Tick the pairs you choose.

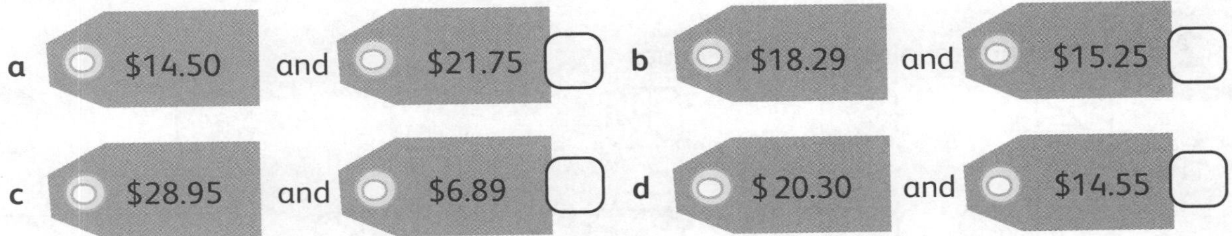

More about sequences

1 Find the rule and the missing terms in these linear sequences.

a ______, $\frac{11}{3}$, ______, ______, $\frac{2}{3}$, ______, $-\frac{4}{3}$ Rule:

b 2.3, ______, 7.3, ______, ______, ______ Rule:

c –50, ______, ______, ______, 110, ______, ______ Rule:

2  Look at the start of this sequence of numbers.

9, 18, 27, …

a Fill in this table.

Position	6		9			39	99
Term		72		108	180		

b Explain why the value of the 200th term is not 1 791.

____

____

c Explain why the number 3 682 will not be a term in this sequence.

____

____

Cube numbers

1 Tick (✓) the two shapes that represent cube numbers.

a 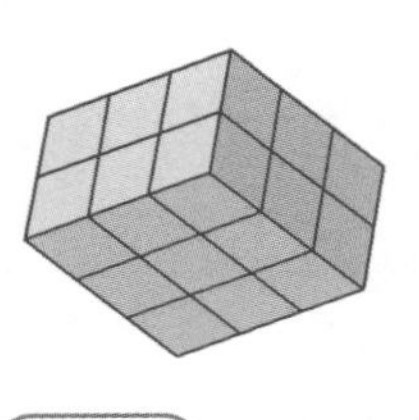

b 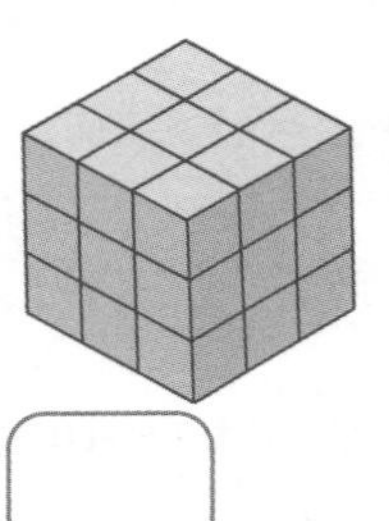

c 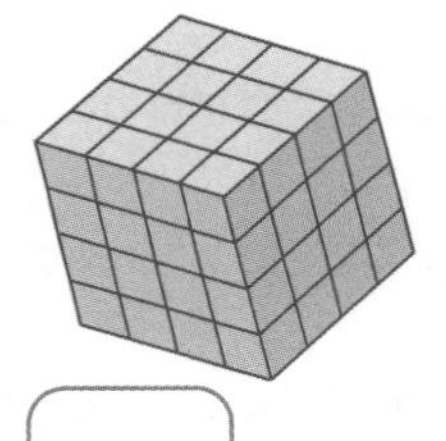

d 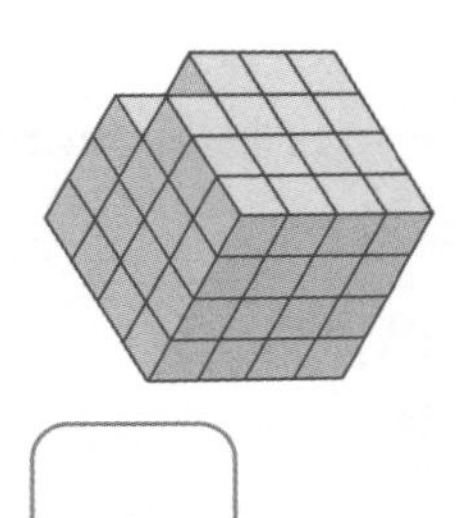

2 Draw lines to join equivalent values.

125 | 27  | 1^3 | 8  | 64 | 25

$2^2 \times 2$ | 4^3 | 5^3 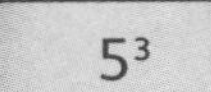 | $5^3 - 10^2$ 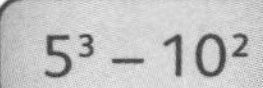 | $3 \times 3 \times 3$ | 1×1^2

3

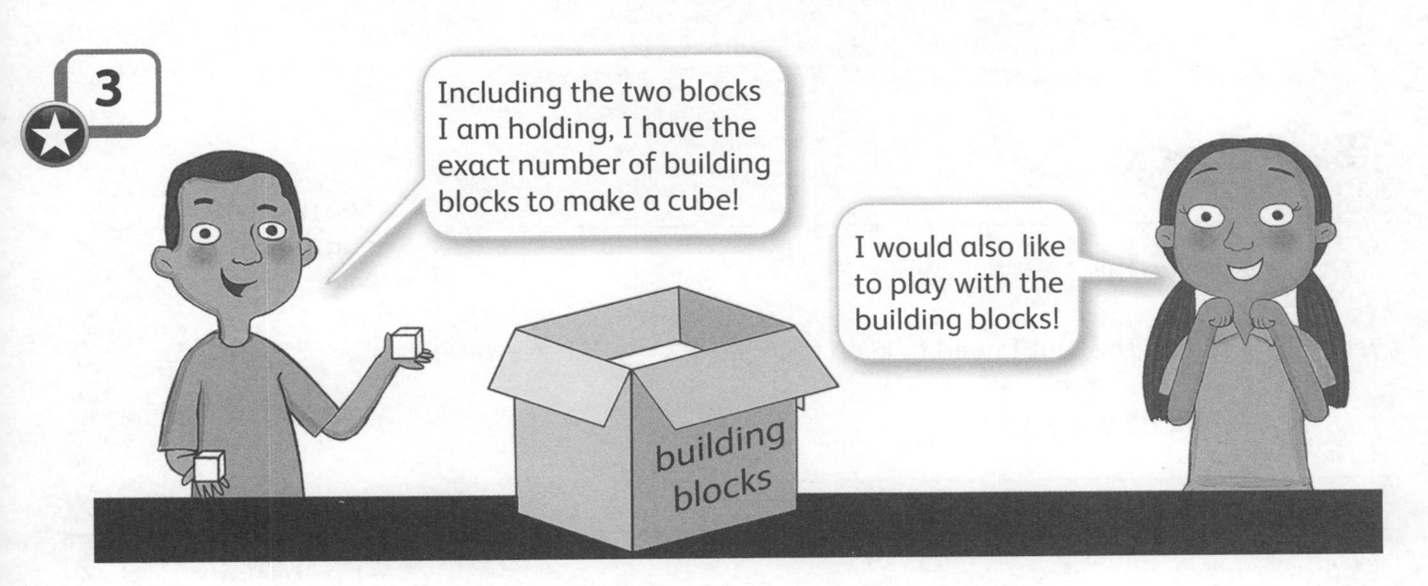

Jin gives Elok 37 of his building blocks.
Luckily, he still has the exact number of blocks to build a smaller cube!
How many blocks did Jin have to begin with? Explain how you know.

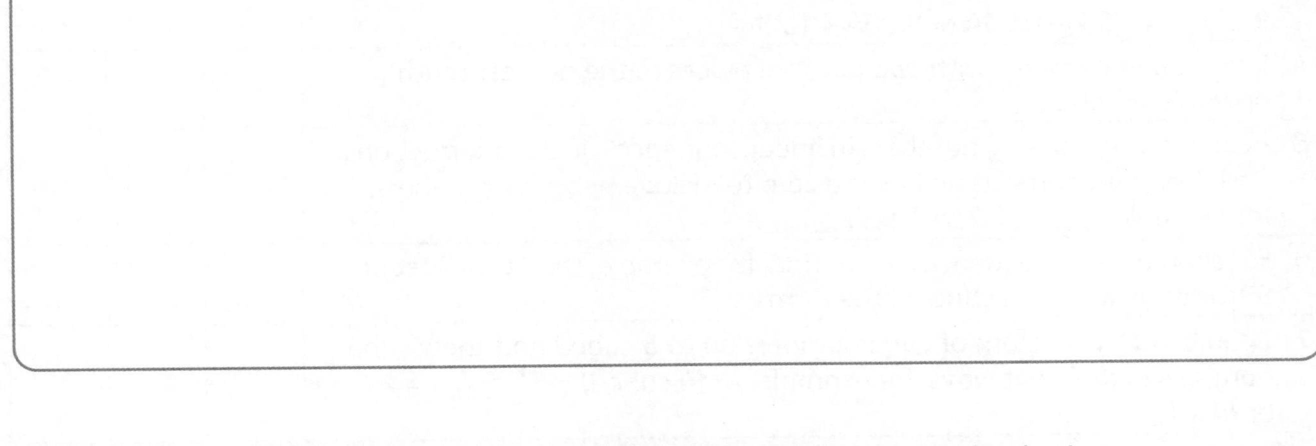

4 Sanchia is thinking of a square number. She multiplies her number by 4.
Now she has a number that is also a cube number!
What number was Sanchia thinking about?
Use the space below to help you work it out.

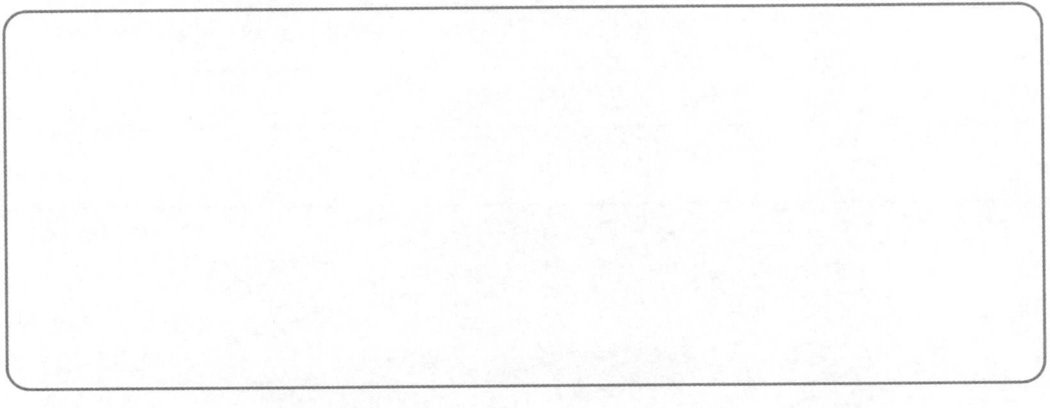

Unit 13 Number

Self-check

See how much you know!

 I can do this.

 I can do this, but I need to keep trying.

 I can't do this yet.

What can I do?	I can do this	I need to keep trying	I can't do this yet
1 I can state and explain the value of each digit in decimals (tenths, hundredths and thousandths).			
2 I can compose and decompose numbers using place value, including decimals (tenths, hundredths and thousandths).			
3 I can regroup numbers, including decimals (tenths, hundredths and thousandths) in a variety of ways, for example: 9.234 = 4 + 5.2 + 0.034 or 8.004 + 1.23.			
4 I can round numbers with two decimal places to the nearest tenth or whole number.			
5 I can identify missing numbers in linear sequences, including fractions and decimals, and extend beyond zero to include negative numbers, for example: __, __, 2.7, 6.2, 9.7, __, __.			
6 For simple linear sequences, I can find, for example, the 10th, 31st or 99th terms without listing all the terms.			
7 I can build the pattern of cube numbers up to 5 cubed and show cube numbers in different ways, for example: 4^3 (4 cubed) = 64 or 4 × 4 × 4, or 4^2 × 4.			

I need more help with:

____

____

____

Unit 14 The coordinate grid

Can you remember?

Plot these points on the grid.
(−5, 5)
(−3, 3)
(2, −2)
(4, −4)

Draw a line through all the points.

Write the coordinates of four other points that are on the line you drew.

____

____

____

____

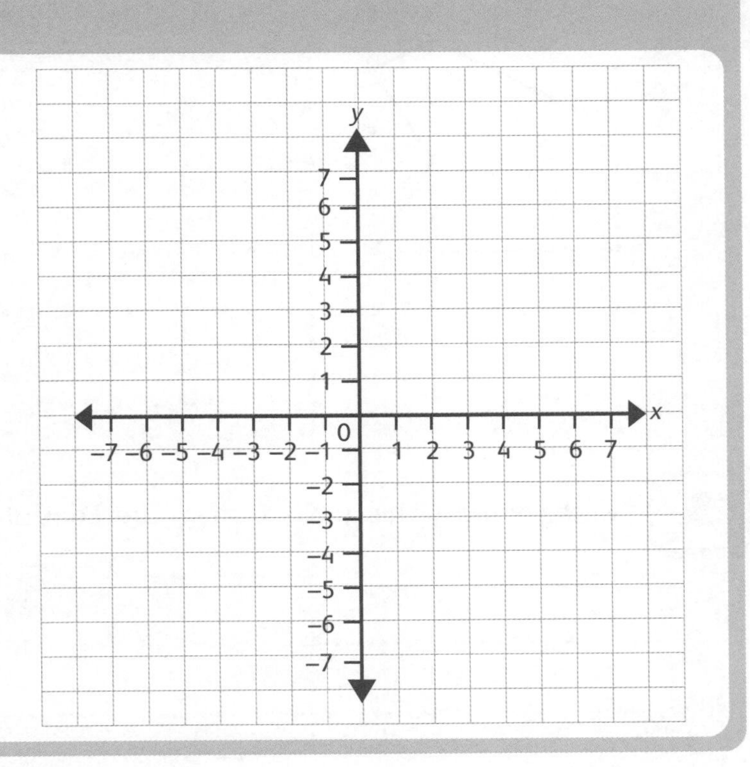

Reflections

1 Complete each reflection.

a

b

c

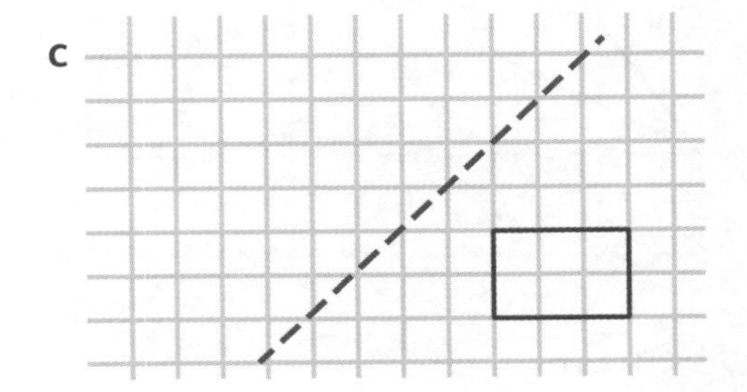

d

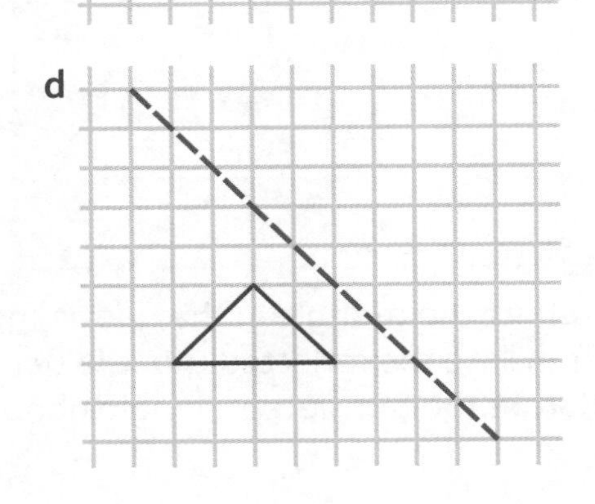

2 Complete each of these reflections.

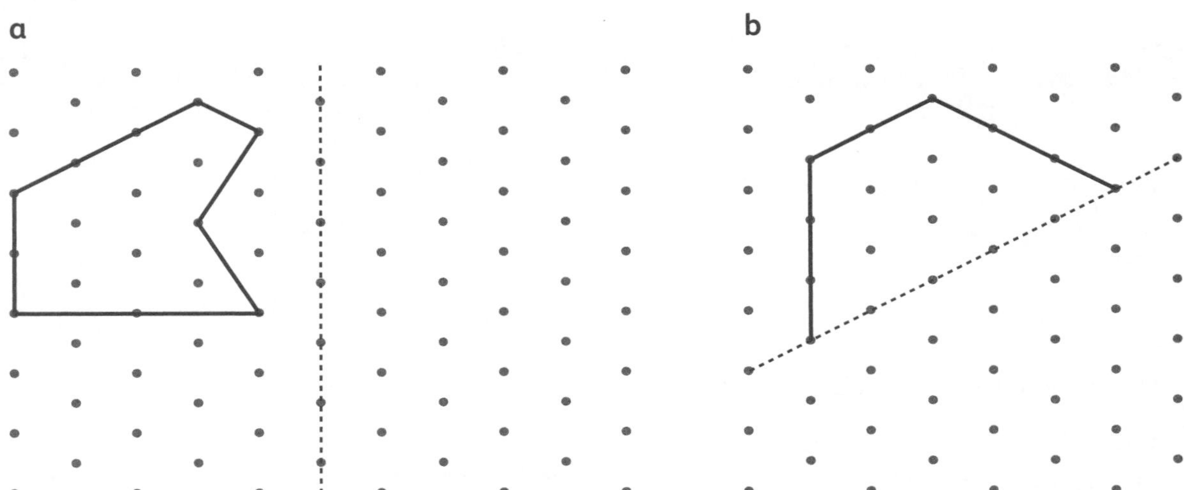

3 Each square in this grid is 1 cm square. Look at triangle ABC.

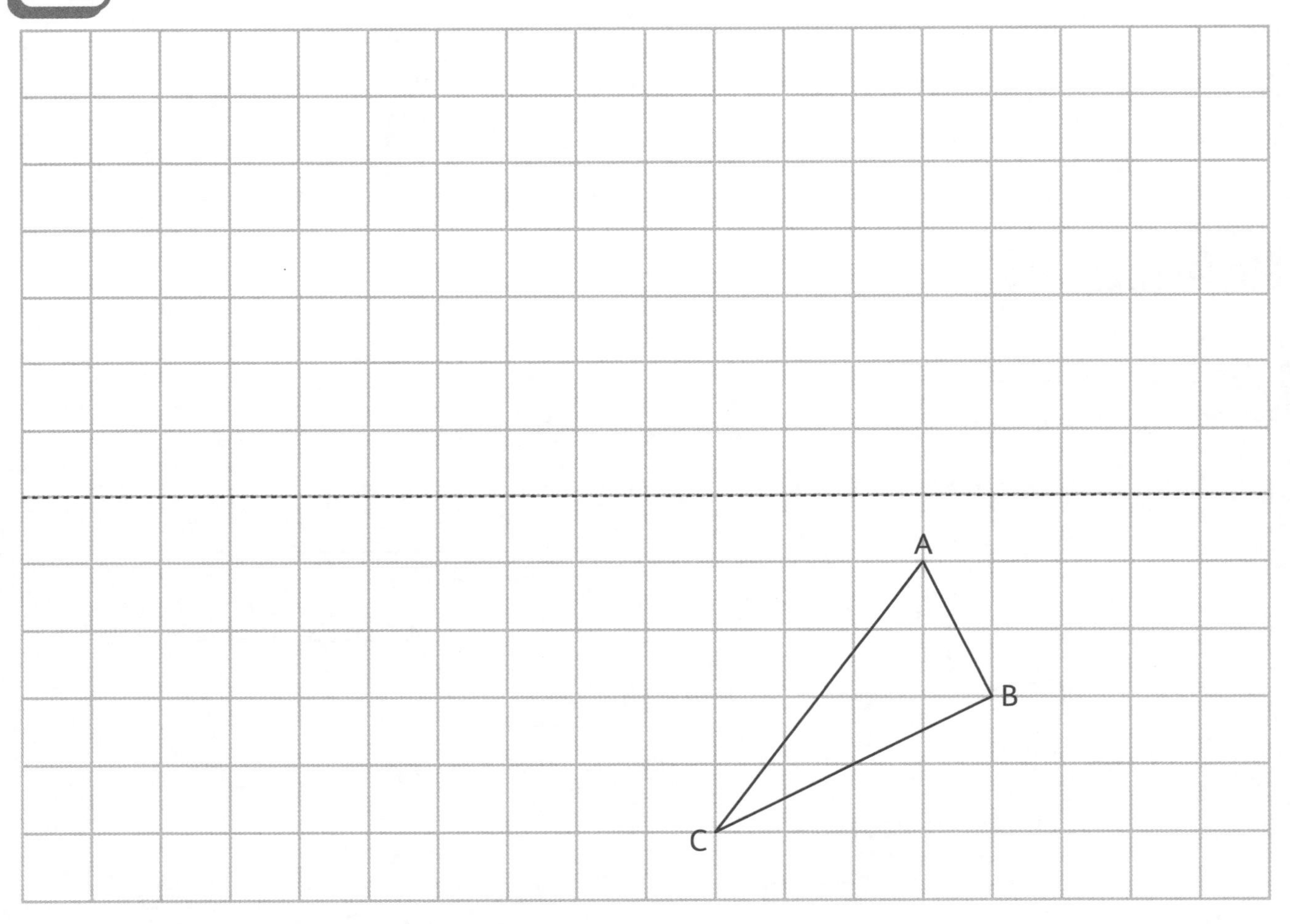

Reflect the triangle in the mirror line.
Translate the reflected triangle 6 cm left and 8 cm up.
Rotate the translated triangle about one of its vertices by 90 degrees.

4

Reflect the pattern in the mirror line.

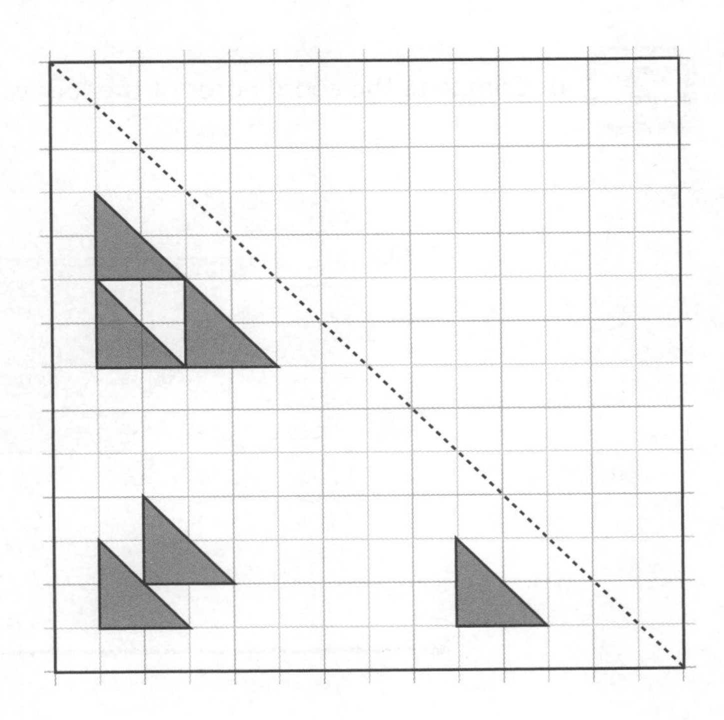

Coordinates

1

a Look at the points on the coordinate grid below. Fill in each coordinate:

A (____, ____), B (____, ____), C (____, ____) and D (____, ____).

b On the same coordinate grid, mark the points below clearly.

P (–1.5, –1.5)

Q ($4\frac{1}{2}, -4\frac{1}{2}$)

R (–4.5, 4.5)

S (1.25, 4.75)

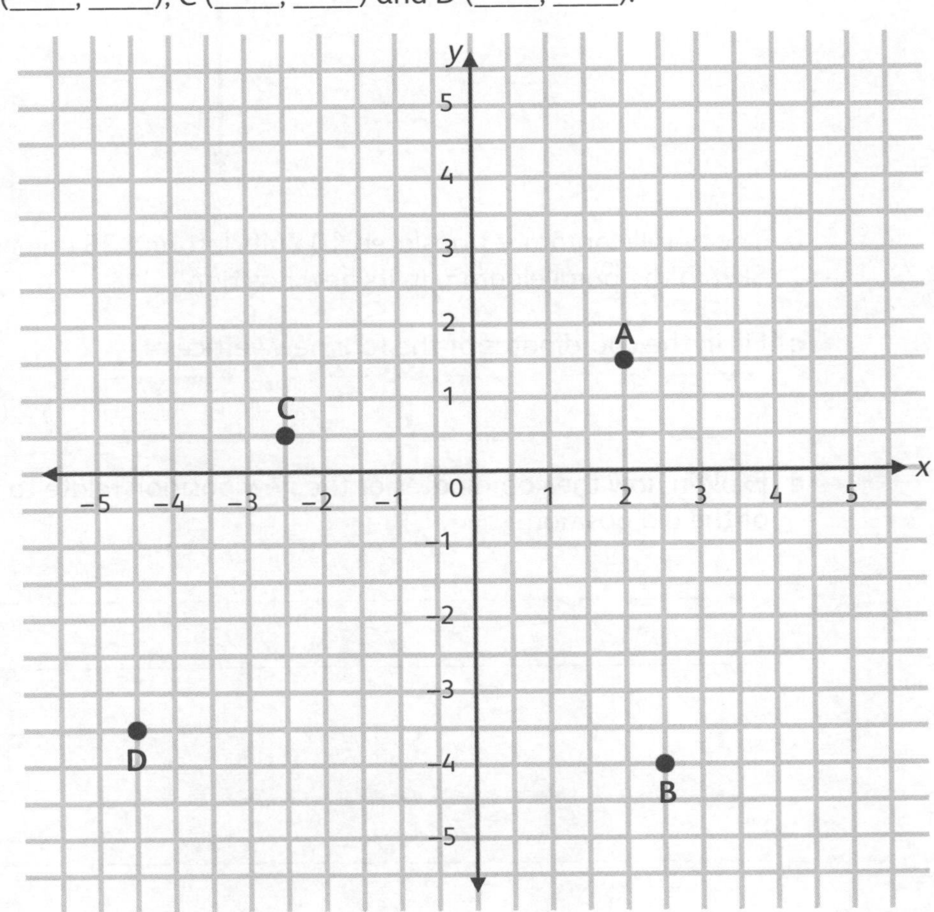

2 **a** Complete the coordinates of the four vertices of the parallelogram.

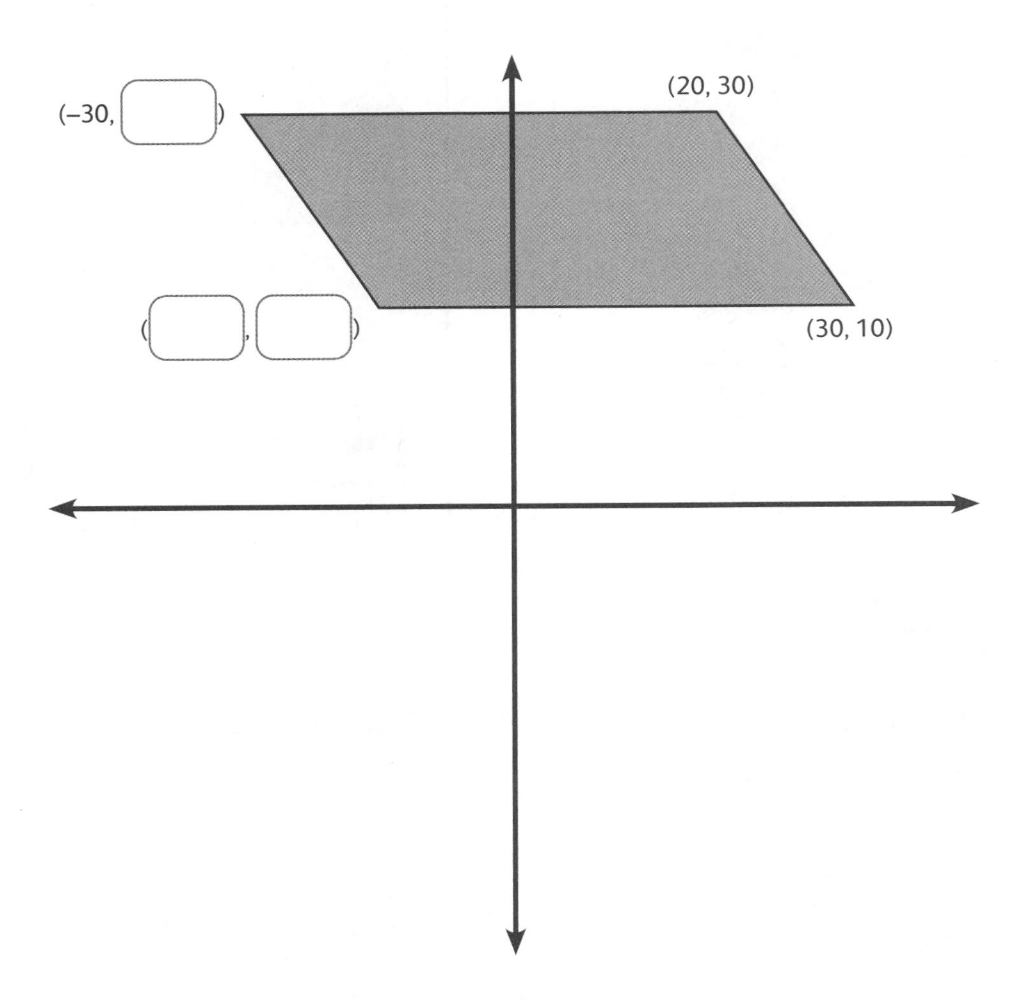

b The parallelogram is translated 10 units left and 25 units down.
Sketch the parallelogram in its new position.

c Fill in the coordinates of the four new vertices:

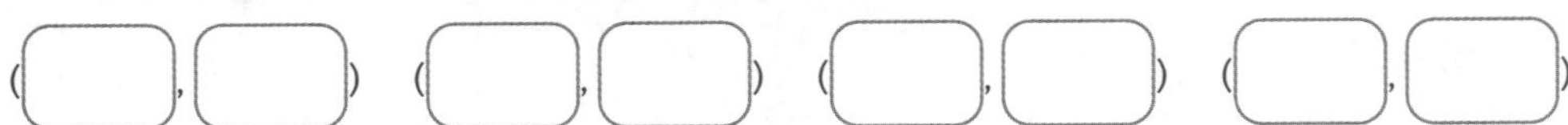

d Explain how the coordinates of the new position relate to the coordinates of the old position.

Unit 14 The coordinate grid

Self-check

 I can do this.

 I can do this, but I need to keep trying.

 I can't do this yet.

What can I do?	I can do this	I need to keep trying	I can't do this yet
1 I can read coordinates, including integers, fractions and decimals, in all four quadrants (with the help of a grid).			
2 I can plot coordinates, including integers, fractions and decimals, in all four quadrants (with the help of a grid).			
3 I can plot points to form lines and shapes in all four quadrants, drawing on my knowledge of 2D shapes and coordinates.			
4 I can reflect 2D shapes in a given mirror line (vertical, horizontal and diagonal) on square grids.			

I need more help with:

____

____

____

Unit 15 Calculation

Can you remember?

Write the first five numbers for each group:

Square numbers: ☐ ☐ ☐ ☐ ☐

Cube numbers: ☐ ☐ ☐ ☐ ☐

Triangular numbers: ☐ ☐ ☐ ☐ ☐

Prime numbers: ☐ ☐ ☐ ☐ ☐

Common multiples of 3 and 4: ☐ ☐ ☐ ☐ ☐

Adding and subtracting fractions

1 Banko draws diagrams to help him add $\frac{4}{5}$ and $\frac{3}{4}$.

$\frac{1}{5}$	$\frac{1}{5}$	$\frac{1}{5}$	$\frac{1}{5}$	$\frac{1}{5}$

$\frac{1}{4}$	$\frac{1}{4}$	$\frac{1}{4}$	$\frac{1}{4}$

$\frac{12}{15} + \frac{9}{15}$ is equal to $\frac{21}{15}$ or $1\frac{2}{5}$

Critique Banko's work.
Use the space to draw any **improvements** he should make. Find the answer.

2 Complete these calculations.

a $\frac{15}{8} - \frac{2}{3} = \frac{\square}{\square}$ b $\frac{13}{10} - \frac{3}{4} = \frac{\square}{\square}$ c $\frac{2}{3} + \frac{6}{5} = \frac{\square}{\square}$ d $\frac{11}{12} - \frac{2}{3} = \frac{\square}{\square}$

3 Guss measures out $\frac{7}{8}$ kg of cherries.
Sanchia measures out another $\frac{3}{5}$ kg of cherries.
What is the total mass of the cherries? ________________ kg

Adding and subtracting decimal numbers

1 A trampoline has a weight limit of 100 kg.
Is it safe for four children to bounce together if they each weigh:
21.7 kg, 24.5 kg, 19.65 kg and 23.80 kg? Show how you know.

2 A delivery van makes trips of these distances:

3.6 km | 9.98 km | 3.4 km | 9.47 km

How far has the van travelled in total? ________________ km

3 Use the column method to solve these calculations.

a 354.6 + 123.35	b 354.6 – 123.35	c 68.479 + 32.3

4 Look at the map of the town and answer the questions.

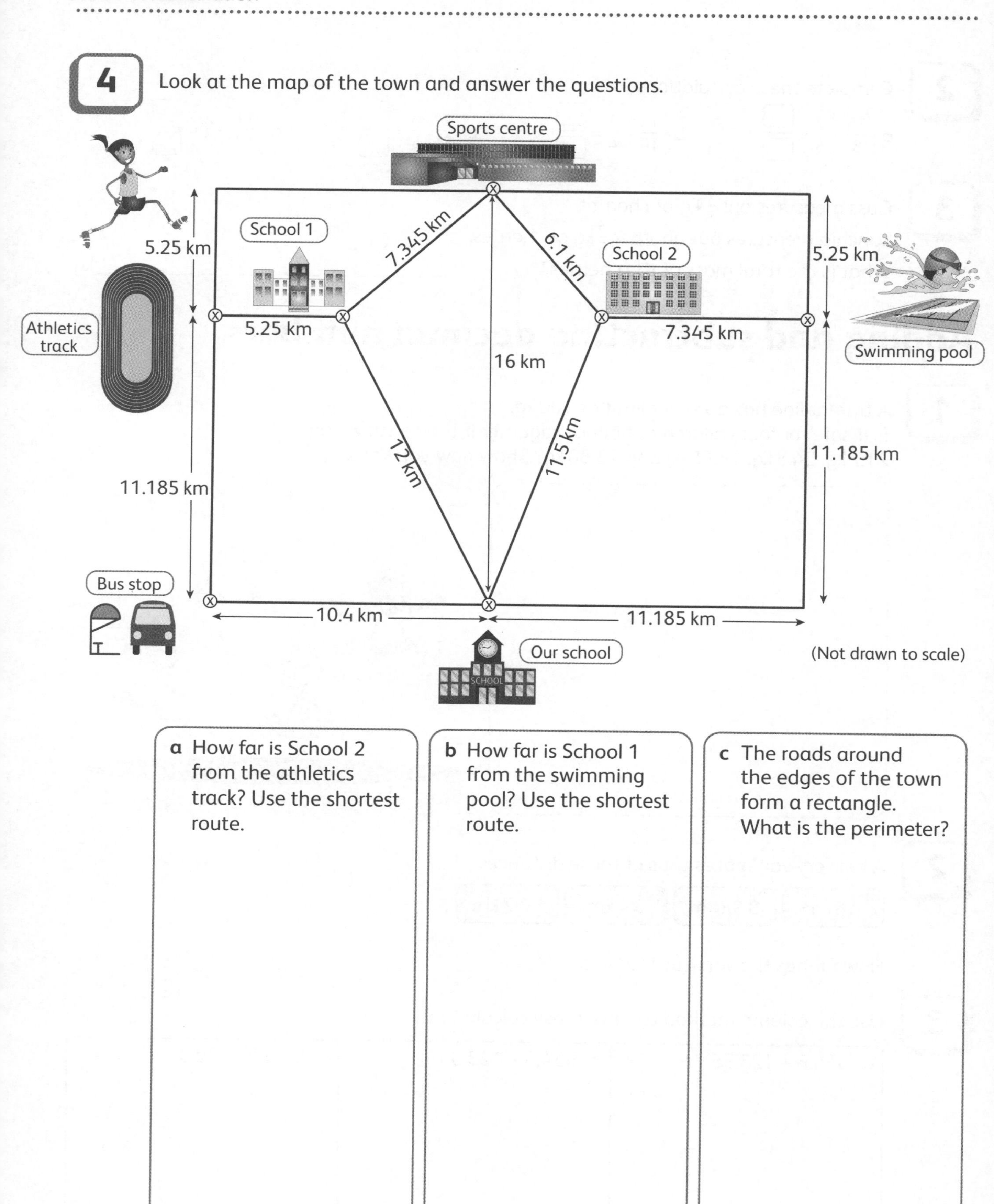

a How far is School 2 from the athletics track? Use the shortest route.

b How far is School 1 from the swimming pool? Use the shortest route.

c The roads around the edges of the town form a rectangle. What is the perimeter?

Multiplying and dividing proper fractions by whole numbers

1 What division calculations are represented here? Solve them.

a

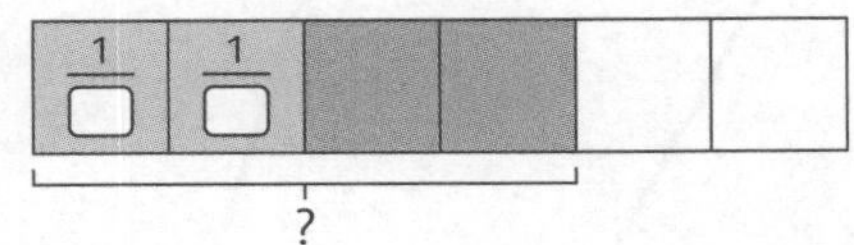

$\frac{\square}{\square} \div \square = \frac{\square}{\square}$

b

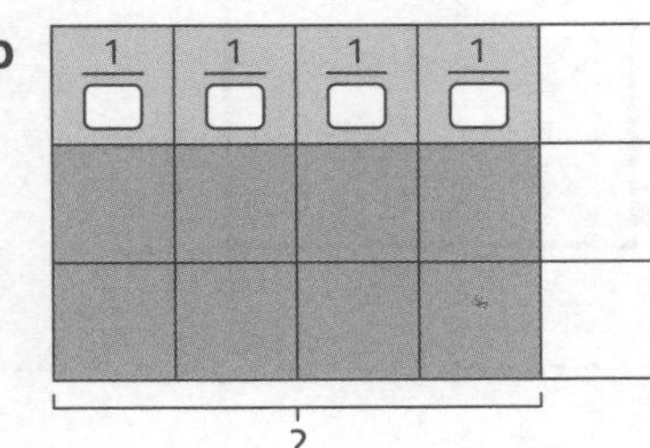

$\frac{\square}{\square} \div \square = \frac{\square}{\square}$

2 What multiplication calculations are represented here? Solve them.

a

$\frac{\square}{\square} \times \square = \frac{\square}{\square}$

b

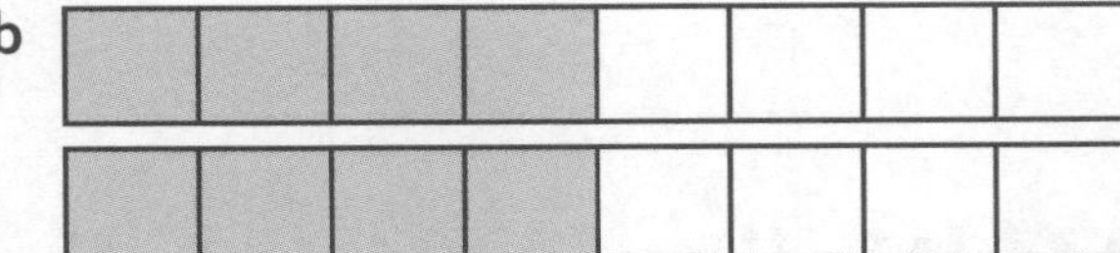

$\frac{\square}{\square} \times \square = \frac{\square}{\square}$

3 Calculate the area of each rectangle. Use the measurements given.

Rectangle	Width	Length	Calculation	Area
A	$\frac{7}{8}$ m	3 m		☐ m^2
B	$\frac{3}{5}$ m	7 m		☐ m^2
C	$\frac{9}{10}$ m	11 m		☐ m^2

4 The perimeter of each of these regular shapes is $\frac{8}{10}$ of a metre.

What is the length of one side of each shape? Write the calculation you use each time.

a ______

b ______

c ______

d ______

e ______

Multiplication and division

1 Complete these divisions.

a $15\overline{)645}$

b $12\overline{)855}$

c $11\overline{)587}$

2 A car hire company buys 16 of the same vehicles. Each vehicle costs $8 979.

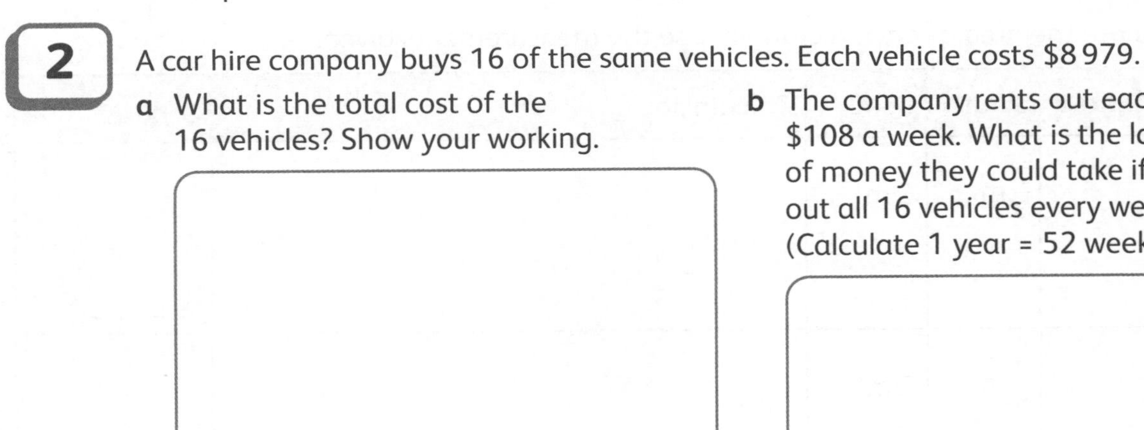

a What is the total cost of the 16 vehicles? Show your working.

$ ______

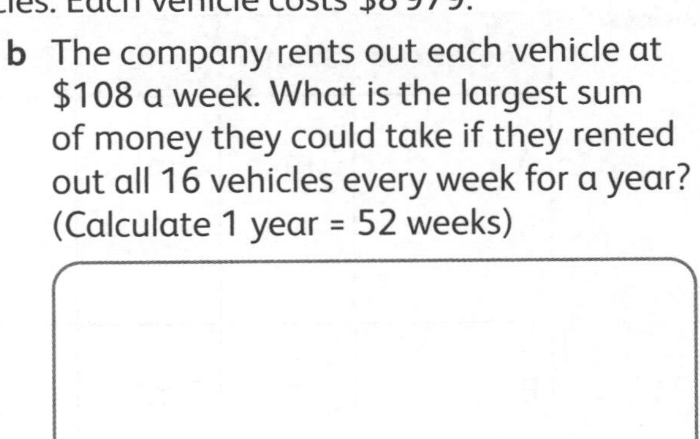

b The company rents out each vehicle at $108 a week. What is the largest sum of money they could take if they rented out all 16 vehicles every week for a year? (Calculate 1 year = 52 weeks)

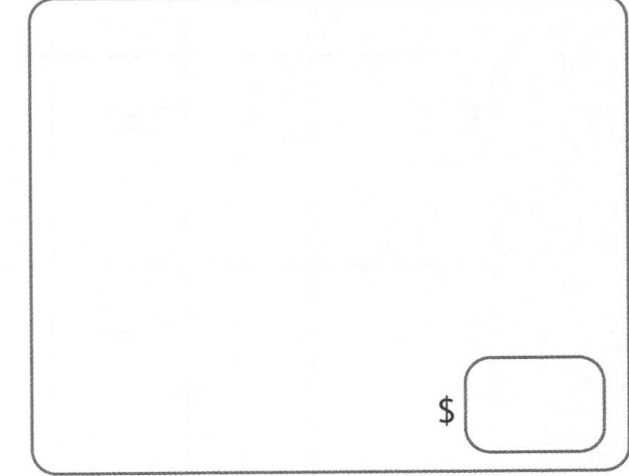

$ ______

3 Decide whether you need to round the answer or turn the remainder into a fraction.

a A box of 16 pens costs $28. What is the cost of one pen? ______

b A path is 635 cm long. How many 50 cm slabs are needed to complete the whole path? ______

c Lemons are packed in boxes of 12. There are 364 lemons. How many full boxes of lemons are packed? ______

Multiplying and dividing decimal numbers

1 Complete these calculations.

a

$4.3 \times 9 = \square$

$2.43 \times 9 = \square$

$24.35 \times 9 = \square$

$24.35 \times 18 = \square$

b

$7.2 \div 8 = \square$

$16.72 \div 8 = \square$

$16.72 \div 16 = \square$

$320.64 \div 32 = \square$

c

$\square = 0.07 \times 5$

$\square \div 5 = 0.07$

$\square \div 12 = 4.3$

$4.3 \times 24 = \square$

2 It takes a fairground 'big wheel' 36.48 seconds to make one full turn.

a Use the information to complete this table.

Number of turns	3	5	9	12	15	21
Total time (seconds)						

b

I **conjecture** that I can calculate how long the 'big wheel' takes to make one-quarter of a turn by dividing 36.48 seconds by 4.

Do you agree with Sanchia? Explain how you know.

______

______

c How can you calculate how long it takes the 'big wheel' to make one-third of a turn?

Explain and then calculate. ______

______

Circle the larger amount on each shelf. **Convince** a partner that you are correct. Use the space below each picture to show your calculations.

Hint

1 lb = 0.45 kg
1 kg = 2.2 lb
1 pint = 0.57 litres
1 ℓ = 1.76 pints
You might not need to do the exact conversion.

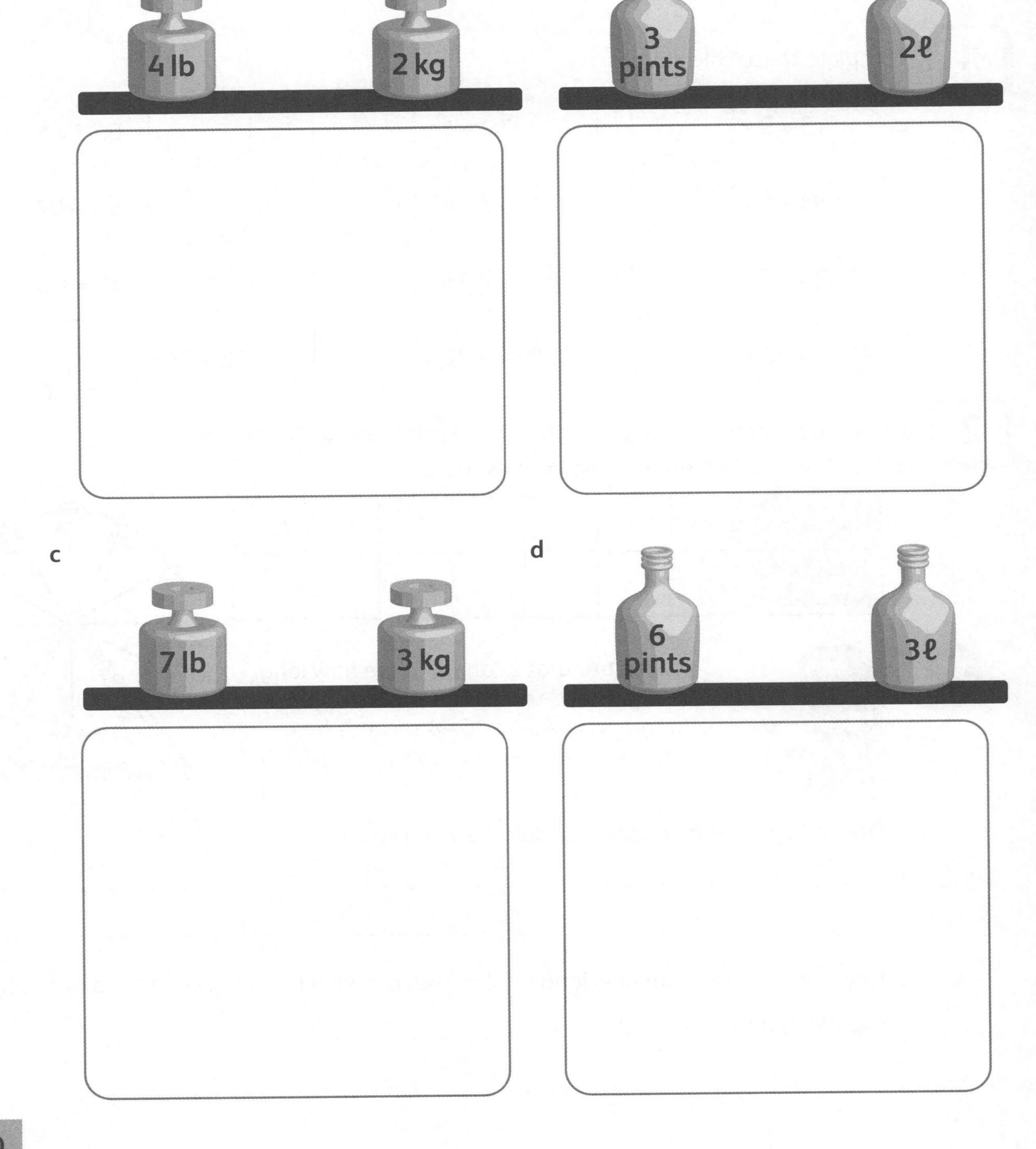

Unit 15 Calculation

Self-check

 I can do this.

 I can do this, but I need to keep trying.

 I can't do this yet.

What can I do?			
1 I can estimate the sums and differences between fractions with different denominators.			
2 I can add and subtract fractions with different denominators.			
3 I can estimate the sums or differences between numbers with the same or different numbers of decimal places.			
4 I can use knowledge of place value to add and subtract numbers with the same or different numbers of decimal places, up to thousandths.			
5 I can explain how using brackets can change the order of operations.			
6 I can estimate the result of multiplying and dividing proper fractions by whole numbers.			
7 I can multiply and divide proper fractions by whole numbers.			
8 I can estimate the product of a whole number up to 10 000 with 1-digit or 2-digit whole numbers.			
9 I can multiply whole numbers up to 10 000 by 1- or 2-digit whole numbers.			
10 I can estimate the result of dividing whole numbers up to 1 000 by 1-digit or 2-digit whole numbers.			
11 I can make an estimate and then divide whole numbers up to 1 000 by 1-digit or 2-digit whole numbers.			
12 I can explain the difference between giving an answer with a remainder or turning the remainder into a fraction.			
13 I can apply the laws of arithmetic and the order of operations to simplify calculations.			
14 I can make estimates and then multiply or divide numbers with one or two decimal places by 1-digit or 2-digit whole numbers.			

I need more help with:

____

____

____

Unit 16 2D and 3D shapes

Can you remember?

This equilateral triangle has a perimeter of 48 cm. → (Not drawn to scale)

Use this information to calculate the perimeter of each shape below.
(Not drawn to scale)

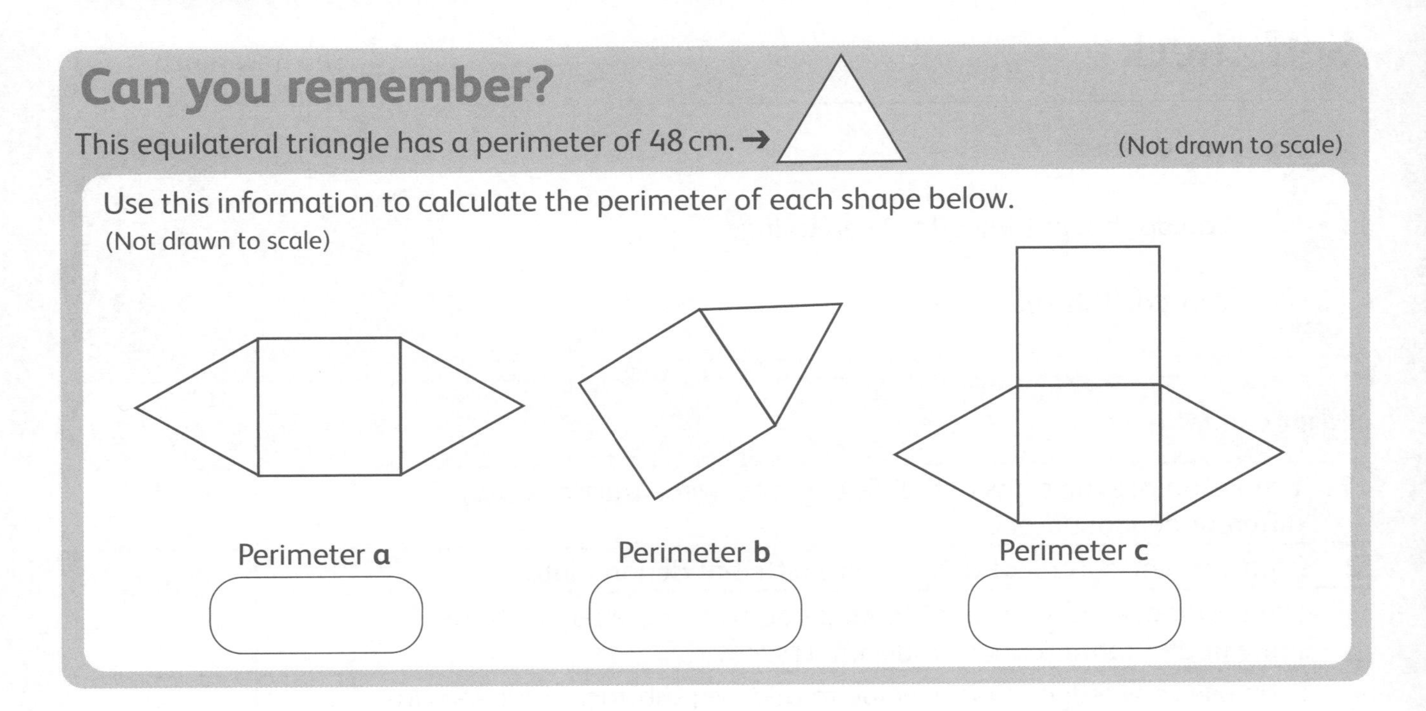

Area

1 Predict which triangles will have an area of less than 10 cm^2.
Then measure and calculate to check.
I predict that ______________________ will have an area of less than 10 cm^2.

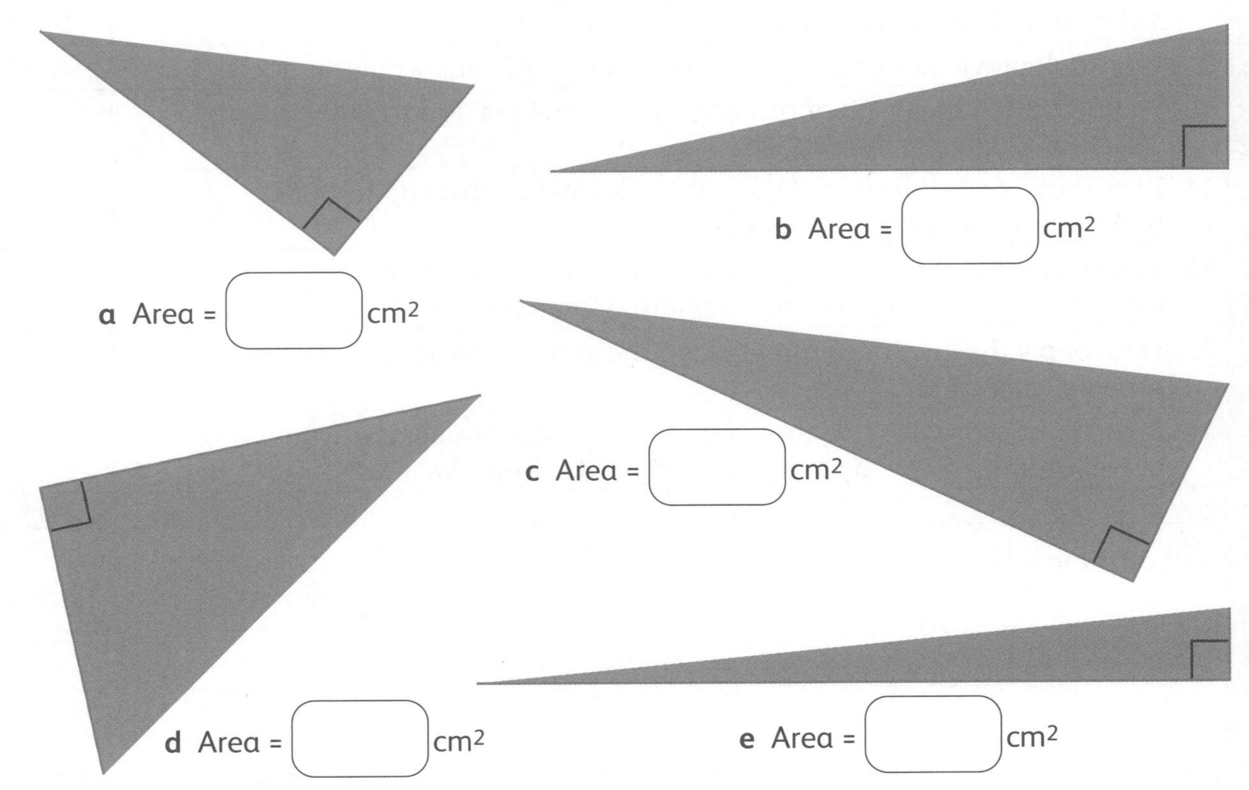

2 Draw two different triangles, each with an area of 15 cm^2.

3 The length of each side of a cube is 4 cm. Sketch a net of this cube.
Then calculate the total surface area. **Convince** a partner that you are correct.

4 Sketch a net for a cuboid that is not a cube. It must have a total surface area of 40 cm^2.

Circles

1 Draw lines to match the words to the correct circles.

circumference | centre | diameter | radius

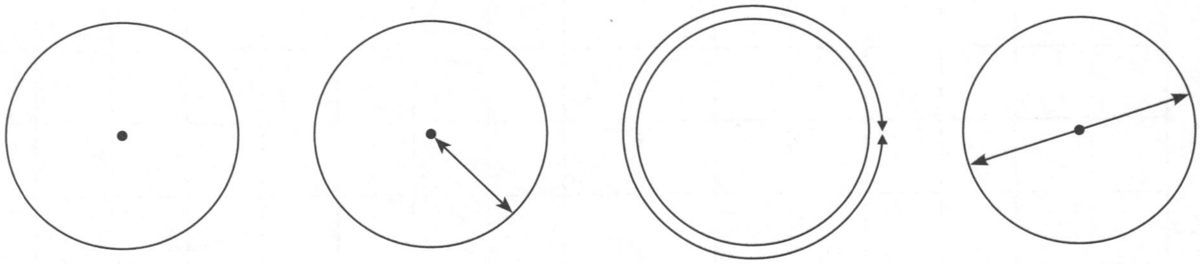

2 Measure the radius and diameter of each circle.

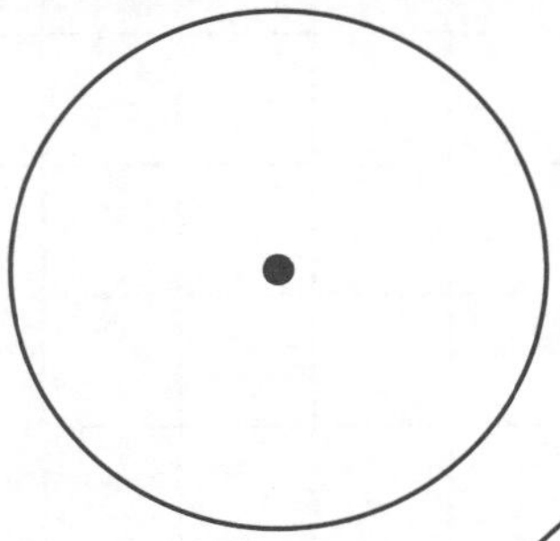

radius

diameter

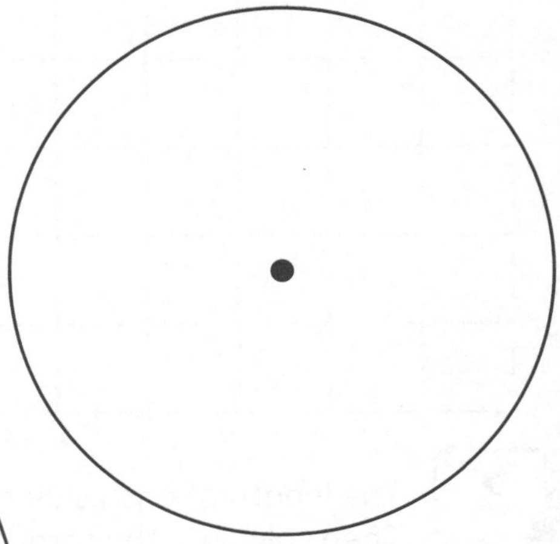

radius

diameter

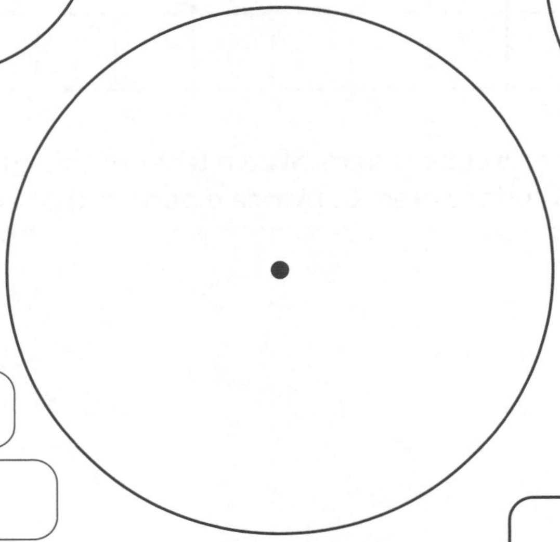

radius

diameter

3 Draw 20 points, each exactly 25 mm from the dot marked C. Use a pair of compasses to join the dots to form a circle.
Write the radius and diameter of the circle.

Radius

Diameter

C

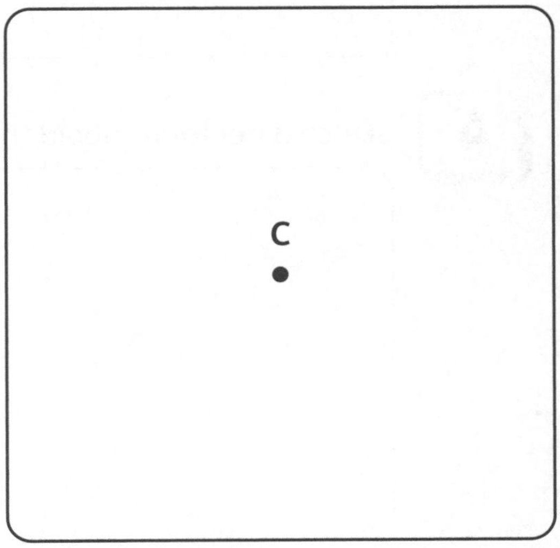

Rotational symmetry

1 Sketch a shape for each order of rotational symmetry.

Order-2 rotational symmetry	Order-3 rotational symmetry	Order-4 rotational symmetry

2 Look at the example.
Then draw a shape for each section of the Carroll diagram.

	Reflective symmetry	No reflective symmetry
Rotational symmetry		
No rotational symmetry		

3 Colour in more squares so that each pattern has rotational symmetry of order 2.

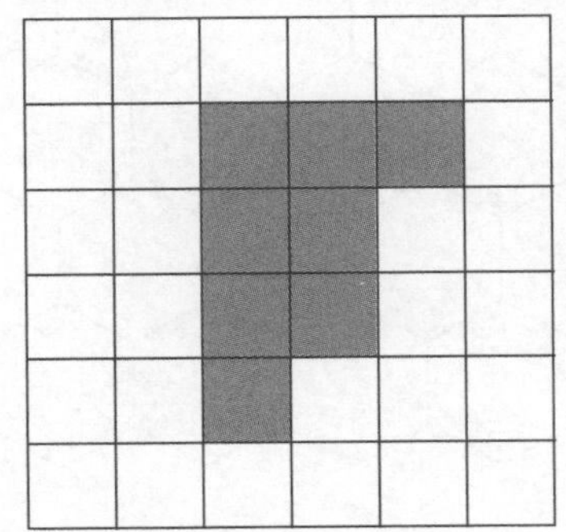

Use the grids below to design shapes and patterns with different orders of rotational symmetry. Label each shape with its order of rotational symmetry.

Unit 16 2D and 3D shapes

Self-check

See how much you know!

 I can do this.

 I can do this, but I need to keep trying.

 I can't do this yet.

What can I do?			
1 I can estimate and calculate the area of right-angled triangles using my knowledge of the area of rectangles.			
2 I can find the surface area of 3D shapes.			
3 I can identify and name the parts of a circle (centre, radius, diameter and circumference).			
4 I can identify rotational symmetry in familiar shapes, patterns or images.			
5 I can describe rotational symmetry.			
6 I can construct circles of a specified radius or diameter.			

I need more help with:

____

____

____

Unit 17 Fractions, decimals, percentages, ratio and proportion

Can you remember?

Computers sometimes use 'progress bars' like this one. It shows how much of a task the computer has completed.

a What fraction of the work has the computer completed? (Do not forget to simplify the fraction to its lowest terms.)

b What percentage of the work has the computer completed?

c It has taken 90 seconds for the computer to get this far. How much longer will it take to finish?

Comparing and ordering decimal numbers

1 Use the digits 0, 1 and 2 each time to make these statements true.

a $\square.\square\square < 0.2$

b $\square.\square > \square.5$

c $1.\square\square < 1.\square$

d $\square.3 > \square.9 > 0.\square$

2 David and Maris record the lengths of five pencils and show them on a chart.

14.2 cm | 12.3 cm | 12.75 cm | 11.7 cm | 13.25 cm

Write the correct measurement next to each pencil.

(Not drawn to scale)

Fractions, decimals and percentages

1

What could Banko's fraction be? Find all the possibilities.

____

2 Write the quantities in order from largest to smallest.

a $0.4, \frac{3}{6}, 75\,\%, \frac{60}{100}$ ______________________________

Largest Smallest

b 0.6, 35 %, 0.45, 6 % ______________________________

Largest Smallest

3 Look at these fractions and mixed numbers.

$\frac{19}{24}$ $\frac{12}{24}$ $2\frac{4}{5}$ $2\frac{9}{15}$ $1\frac{50}{100}$ $\frac{9}{10}$

a Circle the fractions and mixed numbers that are written in their simplest form.

b Write the others in their simplest forms. ______________________________

c Write the fraction or mixed number from above that will make these statements true.

☐ < 0.7 ☐ = 2.8

Calculating with percentages

1 Draw two different shapes. Shade 45 % each time.

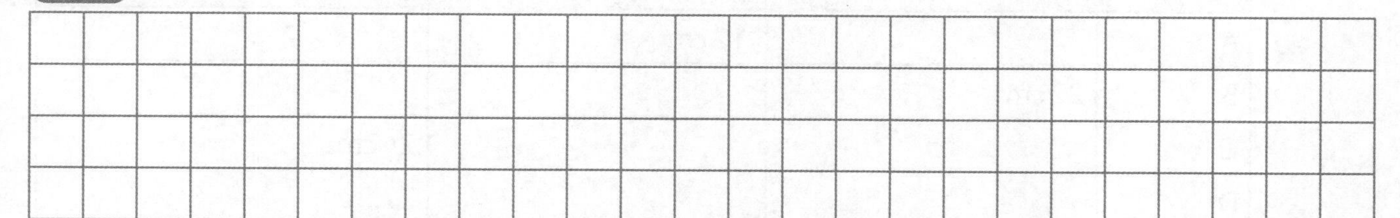

2 Spin a 1 to 6 spinner and choose one of the calculations below. Fill in the percentage and complete the calculation. Repeat until you have filled in all the boxes.

Spin a 1	Spin a 2	Spin a 3	Spin a 4	Spin a 5	Spin a 6
Find 80 %	Find 20 %	Find 35 %	Find 5 %	Find 95 %	You choose

☐ % of $30 = $☐ ☐ % of $60 = $☐

☐ % of $75 = $☐ ☐ % of $150 = $☐

☐ % of $ 800 = $☐ ☐ % of $1 600 = $☐

☐ % of $3 = $☐ ☐ % of $6 = $☐

3 The table shows the number of members at three clubs.

Club	Sports	Art	Music
Number of members	160	80	120

The next year, all three clubs recorded 15 % more members.
How many members did each club have then?

Sports: ☐ Art: ☐ Music: ☐

More about direct proportion

1 Look at the dimensions of each shape.

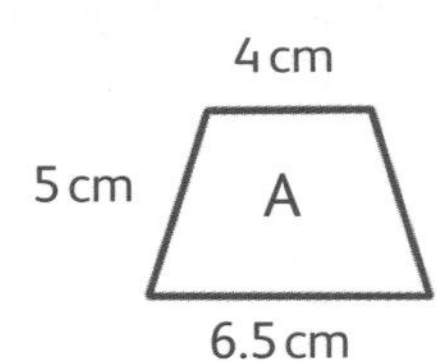

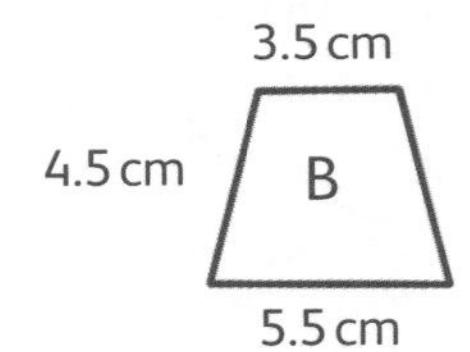

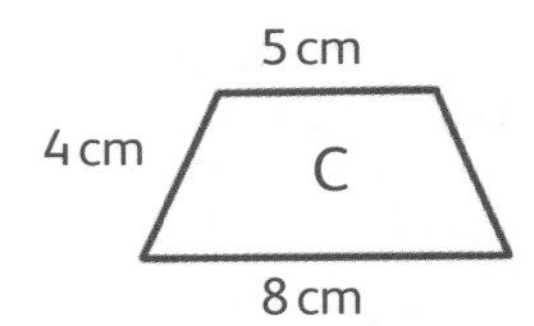

4 cm
4.5 cm
D
7 cm

Complete this table to show the dimensions of the shapes so that they are in proportion to each shape above.

Shape	Base of shape	Top of shape	Sloped sides of shape
A		12 cm	
B	22 cm		
C			24 cm
D			22.5 cm

2 Sanchia and Jin draw a plan for their model castle.
The lengths in the real model castle will be 15 times as long.

a Use a ruler to measure the dimensions accurately.
Write them on the castle plan.

b Now work out and write the dimensions for the real model castle.

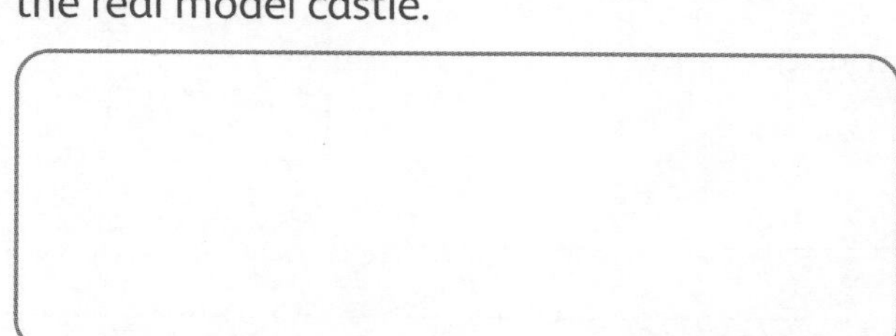

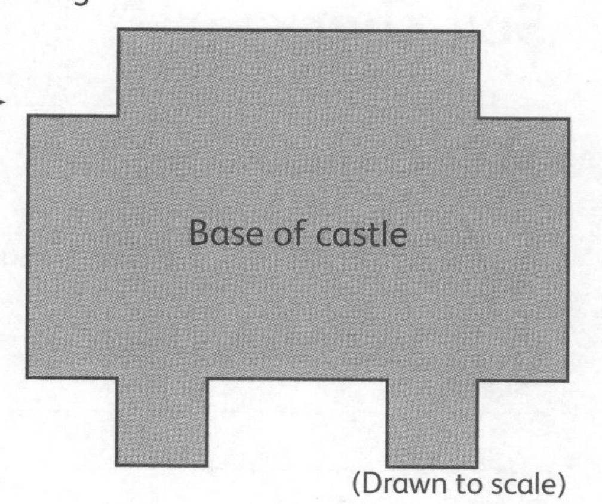

(Drawn to scale)

Ratio problems

1 Use two different colours. Shade triangles to make three shapes with ratios of colours that are equivalent to those in the shape below.

2 Complete these equivalent ratio diagrams.

a

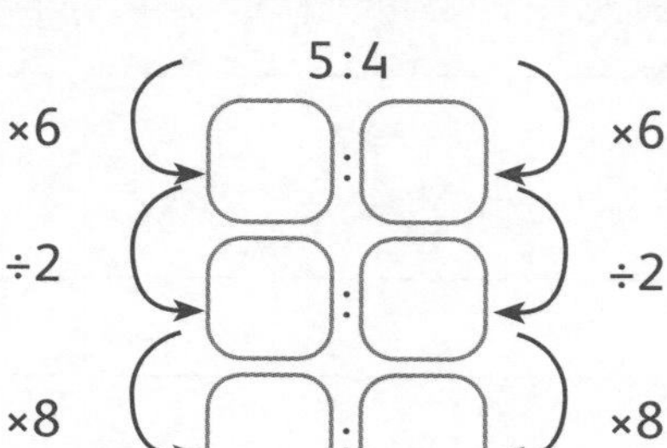

b

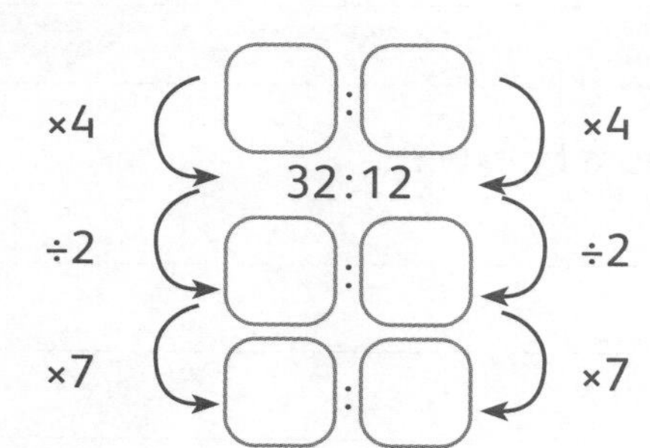

3 Banko and Pia planted some seeds in the garden.
For every five seeds that grew into plants, three seeds did not grow.

a If 35 seeds grew, how many seeds did not grow?

b How many seeds did the children plant in total?

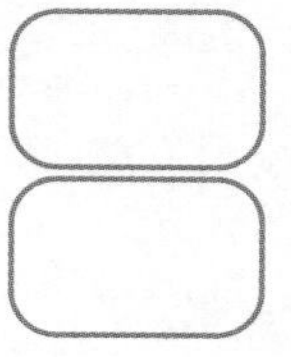

Unit 17 Fractions, decimals, percentages, ratio and proportion

Self-check

 I can do this.

 I can do this, but I need to keep trying.

 I can't do this yet.

What can I do?			
1 I can compare and order numbers with one or two decimal places, using the symbols =, > and <.			
2 I know and can give benchmark equivalences of fractions, decimals (one or two decimal places) and percentages.			
3 I can compare and order numbers with one or two decimal places, proper fractions with different denominators and percentages, using the symbols =, > and <.			
4 I can use my knowledge of equivalence to write fractions in their simplest form.			
5 I can calculate percentages (in multiples of 5) of quantities, for example: Find 15 % of 4 kg.			
6 I can show on diagrams, percentages (in multiples of 5) of shapes, for example: Shade 35 % of a rectangle.			
7 I can give values of measures that are in proportion for simple numbers and in context.			
8 I can solve context problems involving simple ratios.			

I need more help with:

____

____

____

Unit 18 Statistical methods

Can you remember?

Look at the table. Write **true** or **false** for each statement.

Shop	Bicycles sold
A	24
B	18
C	8

Shop A sold three times as many bicycles as Shop C. ________

Shop B sold more than twice as many than Shop C. ________

Shop A sold over 50 % of the total number of bicycles sold. ________

Proportion of the whole

a Colour in the pie charts to represent each set of data about the types of fuel used by different cars in a car wash.

Town Car Wash	
Engine type	**Frequency**
Diesel	9
Electric	3
Petrol	8

City Car Wash	
Engine type	**Frequency**
Diesel	5
Electric	3
Petrol	8

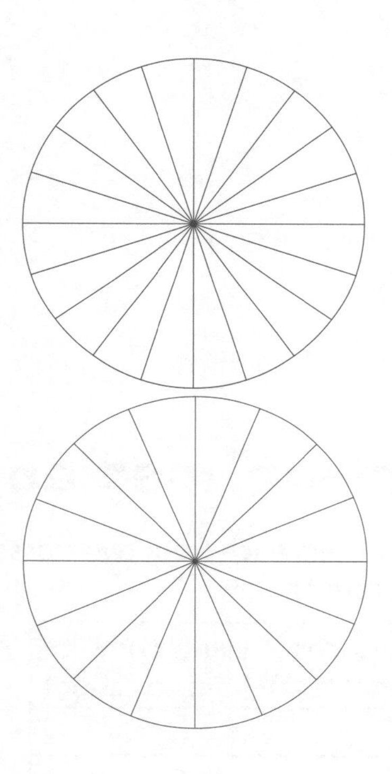

b Which car wash had the greater proportion of petrol cars? ____________

____

Read this probability experiment.
David expects his chance of choosing a black cube, without looking, to be 50 %.
Maris says that the probability of choosing a grey cube should be less than 25 %.
David suggests that they use the spinner to test these predictions.

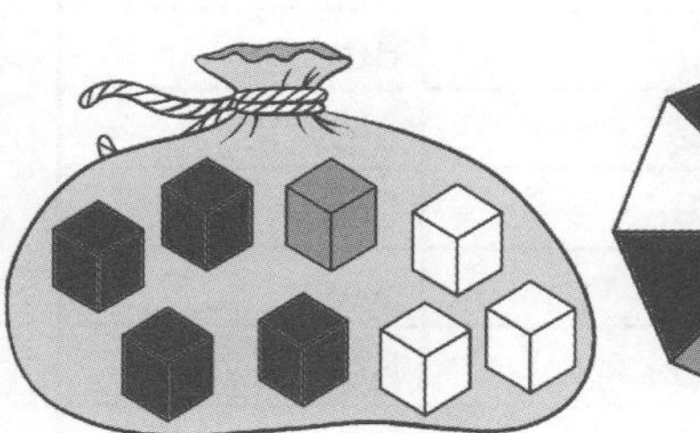

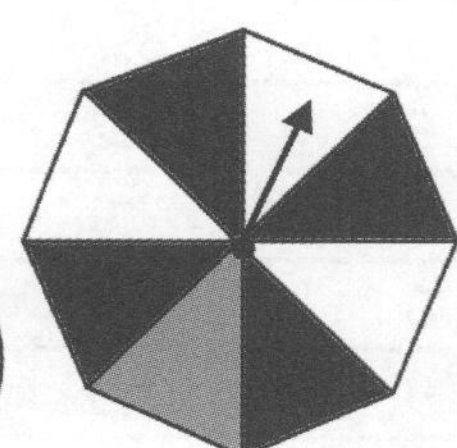

a Complete a probability experiment to test the predictions.
After 4 trials, colour in pie chart A. After 12 trials, colour in pie chart B.
After 20 trials, colour in pie chart C.

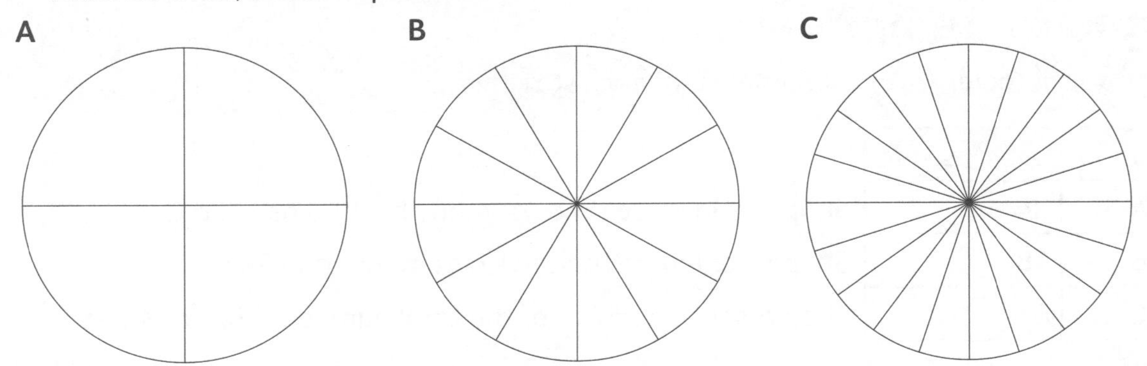

b Sketch a pie chart for what you would expect after 100 trials.

Mode, mean, median, range

Pia asked 10 learners each from two classes about how they get to school.
Her results are listed below.

Learner	Distance to school (km)	How they get to school
Jin's class		
1	2.1	Walk
2	1.2	Walk
3	1.3	Bicycle
4	8.9	Car
5	8.2	Bus
6	1.2	Bicycle
7	3.9	Car
8	2.3	Bus
9	2.3	Bus
10	1.9	Bicycle

Learner	Distance to school (km)	How they get to school
Elok's class		
1	2.6	Car
2	6.1	Bus
3	1.1	Walk
4	4.5	Car
5	15.3	Bus
6	2.3	Bus
7	0.5	Walk
8	2.3	Bus
9	6.1	Bus
10	2.3	Bus

1 a Use the information in Pia's tables to complete these frequency tables.

Jin's class	
Mode of transport	**Tally**
Walk	
Bicycle	
Bus	
Car	
Total (check if it is 10)	

Elok's class	
Mode of transport	**Tally**
Walk	
Bicycle	
Bus	
Car	
Total (check if it is 10)	

b What distance is the mode for each class? Use the box for your calculations.

Mode for Jin's class:

Mode for Elok's class:

2 Calculate the mean distance for each class.

Mean for Jin's class:

Mean for Elok's class:

3 Calculate the median and range of each data set.

Median and range for Jin's class:

Median and range for Elok's class:

4 Compare your findings about each class. What interpretations can you make?
Describe how the mean, mode, median and range reflect the modes of transport used.

Unit 18 Statistical methods

Self-check

 I can do this.

 I can do this, but I need to keep trying.

 I can't do this yet.

What can I do?	I can do this.	I can do this, but I need to keep trying.	I can't do this yet.
1 I can predict outcomes of an investigation.			
2 I can find the mode, median, range and mean of a data set.			
3 I can represent data in pie charts to show proportions of the whole data set.			
4 I can interpret data and draw conclusions.			
5 I can compare data and draw conclusions.			

I need more help with:

____

____

____